CONTEMPORARY ENAMELING

ART AND TECHNIQUES

LILYAN BACHRACH

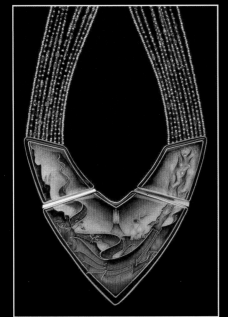

DEDICATION

I dedicated the first edition to my supportive husband, Samuel Bachrach, M.D.
I dedicate this edition to my family of two daughters, two sons, their spouses and off-spring who include eleven grandchildren, three great-grandchildren, and also the many friends who have been so encouraging and helpful.

The Editorial Team for *Enameling With Professionals*
Elizabeth Bachrach Tan, PhD
Barbara Bachrach Scolnick, MS
Martin Communications
Benjamin I. Bachrach, PhD
Robert Z. Bachrach, PhD
Laura Ruth Scolnick, PhD

Editorial Assistance for *Contemporary Enameling*
Barbara Bachrach Scolnick, MS

Library of Congress Cataloging-in-Publication Data

Bachrach, Lilyan.
  Contemporary enameling : art and techniques / Lilyan
Bachrach.— 1st ed.
    p. cm.
  ISBN 0-7643-2355-5 (hardcover)
1. Enamel and enameling—Technique. I. Title.

NK5000.B25 2006
738.4—dc22
                                        2005022964

Designed by John P. Cheek
Cover design by Bruce Waters
Type set in Copprpl Goth Bd BT/Dutch801 Rm BT

ISBN: 0-7643-2355-5
Printed in China

Published by Schiffer Publishing Ltd.
4880 Lower Valley Road
Atglen, PA 19310
Phone: (610) 593-1777; Fax: (610) 593-2002
E-mail: Info@schifferbooks.com

For the largest selection of fine reference books on this and related subjects, please visit our web site at
www.schifferbooks.com
We are always looking for people to write books on new and related subjects. If you have an idea for a book please contact us at the above address.

This book may be purchased from the publisher.
Include $3.95 for shipping.
Please try your bookstore first.
You may write for a free catalog.

In Europe, Schiffer books are distributed by
Bushwood Books
6 Marksbury Ave.
Kew Gardens
Surrey TW9 4JF England
Phone: 44 (0) 20 8392-8585; Fax: 44 (0) 20 8392-9876
E-mail: info@bushwoodbooks.co.uk
Free postage in the U.K., Europe; air mail at cost.

FOREWORD

You are now holding *Contemporary Enameling, Art and Techniques*, an enlarged and all-color edition of *Enameling with Professionals*. Due to its popularity in the enameling community, the limited edition of *Enameling with Professionals* became out-of-print in two years. The momentum of the art of enameling is accelerating and demand is increasing. With additional artists describing "how-I-work" and the many additional color illustrations, this book gives even more than the proven best seller.

Lilyan Bachrach, a veteran enamelist of 50 years, studied with Kenneth Bates, Doris Hall and others from the first "Golden Years" of enameling in America. She has carried much of the tradition into the present revival, or the "Second Golden Period" of enameling, which is only a few years old.

Woodrow Carpenter
Founder, The Enamelist Society

PREFACE TO THE SECOND EDITION

When Nancy Schiffer told me that Schiffer Publishing would like to do an all color, enlarged edition of my first book, *Enameling With Professionals,* I was delighted. The book I published in June, 2002, in a limited edition of 1,000 copies, had taken more than three years to plan, edit, and market. The prospect of a continued life for it was rewarding.

This book, now titled *Contemporary Enameling,* has nine new sections and over 300 new color photographs. I am grateful that so many enamelists have contributed. The techniques for enameling are many and varied, and a wide range of personal innovation, expression and character is possible. The techniques in this book are for vitreous enamels, "hard enamels" on metal as opposed to "soft enameling" that uses paints. The photographs of the work of some of the enamelists show that they have not limited themselves to the one technique featured in the text they wrote while others show the variety possible with one technique. Additional work in enameling is in The Gallery. I wish we had been able to include more from all of the enamelists.

Enameling is both an ancient and contemporary art form. I hope that this edition will continue to enlarge the growing interest in and understanding of enameling. I am proud to have received The Enamelist Society's 2005 Woodrow Carpenter Award, "established to honor those who are presently working to benefit the world of enameling."

Lilyan Bachrach
March 2005

PREFACE TO THE FIRST EDITION

This book has been a long time in coming. I first started writing it in the 1970s when a publisher asked me to write a book on enameling, the craft of fusing glass to metal. I said, "yes" because I wanted to write a book that I could recommend to my students. Well, family matters intervened and I could not take the time to finish the writing. All the notes were stored away until almost two years ago when Jean Jenkins asked me to write a "good" book on enameling because there was not one on the market. I remembered my notes. I had asked a few other professional enamelists to write their methods of working. With each one contributing a text on a different technique, the procedure developed from their practical experience, I knew it would be a better and truer book. I decided to try it again and asked other enamelists to write. The response was wonderful and encouraging.

I am grateful that all the contributors to *Enameling with Professionals* were willing to share what they have learned and made the time to write and revise their texts. They have been very cooperative in answering my many questions about the enamels they use and the details of the way their work. Our intent is to tell you how we enamel in the technique that has been our focus, the technique that we really know, and what works for us. As you will note, there are many ways to enamel. Of course, we agree that there is not one right way. We all

think that the way we work today is the best, but we know that tomorrow our method of working could change. Most of us like to experiment, and in the field of enameling, there is always something new to try.

The easiest way to learn enameling is to take a hands-on course or even a short workshop. This book is intended to help you understand and learn enameling either by yourself or as a supplement to other instruction.

The first part of the book describes my workshop, with its equipment as a frame of reference, and then gives details about the metals for enameling, the enamel material, some beginner enameling techniques, and suggestions for combining them. Doris Hall's paisley design is explained because it shows how many different enameling techniques can be combined.

I owe my interest in enameling to Doris Hall, who died in summer 2001. In 1955, I was oil painting and studying silversmithing when I helped form an enameling workshop with nine other women. We called our group Lenox Enamelers. (It survived for five years.) That year, Doris Hall offered an enameling demonstration course and five of us traveled from Worcester, Massachusetts to her Boston studio once a week. She and her husband, Kalman Kubinyi, worked together producing enamels and commissions. After each lesson we would practice in our studio during the week. At

the end of the course, Doris gave me the 6" paisley plate she had made to demonstrate. She also gave us this advice: "You may copy this to learn, but do not sign your name to the piece you make because you did not design it." The advice is still good.

Doris taught herself to enamel and enameled as a painter. For her enamel paintings, she treated her copper piece as a canvas and drew in a dried layer of opalescent crackle to produce an oxidized line, a technique she originated. When the opalescent crackle as a base coat was fired high it looked like a flux base coat. She had had Ferro Corporation make her the opalescent crackle. Her painting was developed with the sifters Kalman made for her. His sifters were made with a wire holding a piece of screen and a shaped wooden handle to a plastic tube. We adapted that design by fusing the screen to a plastic tube and cementing on a handle. I am still using most of those sifters.

The color illustrations show the enamel work of all the enamelists who are sharing their studio methods.

The second part of the book is what you would learn if you were fortunate enough to enroll in workshops with all of these professional enamelists. The intent is to show you how to enamel, not what to make. Although an enamel piece can be produced without any knowledge of art, for an enamel to be a work of art depends on the craftsman's ability as an artist in design, drawing, painting, or graphics. All of these can come into play with enamels. I had been enameling for ten years before I enrolled in the fine arts program of art school.

Yes, enameling is truly the art of the fire. It needs the heat of a kiln or a torch to fuse a layer of vitreous enamel to the metal. Unlike ceramics, the enamel work is fired for only a few minutes and then removed to cool. Addititional layers of enamel are applied and fired. It is thrilling to watch an enamel piece, after it is removed from the kiln, change color and reveal itself as it cools. The reds, when first taken from a kiln, look brown and slowly become their red as the piece cools. Some opaque enamels fired high become transparents, and when fired at a lower temperature are changed back to opaque. We love enamels for their color, brilliance, tactile quality, and the surprises the kiln can give us. We hope that you will find the same delight and pleasure in this wonderful medium, even from your first piece of enamel. Or, if you have this book for the information it offers, we know it will increase your understanding and appreciation of the art of enameling.

Lilyan Bachrach
March 14, 2002

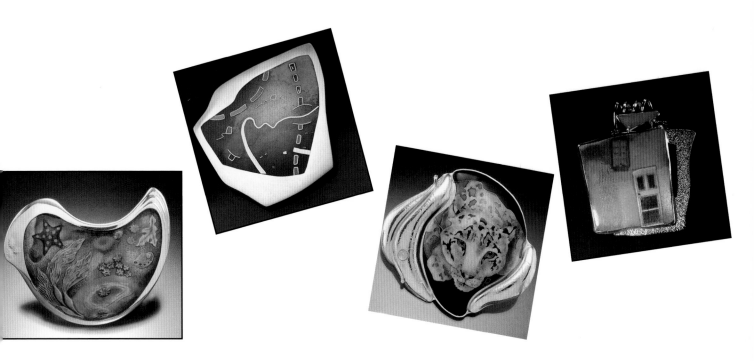

CONTENTS

THE GALLERY
OF CONTEMPORARY ENAMELISTS

INTRODUCTION: ENAMELING BASICS

THE ENAMELIST'S VOCABULARY

Each medium has its own vocabulary and so it is with enamels and enameling. The glass glaze material that is fused to the metal is **enamel**, the finished work is an **enamel piece** and the process is **enameling**. In other words, you enamel an enamel with enamel.

An enamel is **fired** when it is placed in a hot kiln. An electric **kiln** (oven, furnace) is usually used to fuse a **coat** (layer) of enamel to the metal. The pyrometer with the kiln shows the temperature inside the kiln in Fahrenheit (F) or in Centigrade (C). The inside of the kiln is called a **muffle** or **chamber**. The support for transporting the enamel piece in and out of the kiln is a **planche** (firing rack). The tool for lifting the planche is usually a **fork**. A **trivet** or a **hammock** supports an enamel piece by its edges. A **stilt** supports a piece on points in the kiln.

The first layer of enamel on the front and back is a **base coat**. An underfired coat is at the **orange peel stage**. The glossy stage is **at maturity**. **Flash firing** or **healthy firing** is a short, hot firing to quickly gloss the enamel. The size of the enamel grains is the **mesh**. 80 mesh, which is the standard, means that it will sift through a screen that has 80 holes per linear inch. The **fines** are what sifts through after the enamel is sifted through a 325 mesh screen. The enamel material is formulated for firing **soft**, **medium** or **hard**, which refers to the softening point of the enamel. The hardest enamels take the longest to fire to maturity. Methods of applying enamel include **sifting**, **spraying**, **wet packing** (inlaying) and **Indian sand painting**.

The following abbreviations are used in the text: **cm**, centimeter; **ga**, guage; **k** or **kt**, karat (for gold), **mm**, millimeter.■

AN ENAMELING WORKSHOP

Enameling is the process of fusing glass to metal. A layer of glass is applied to a piece of metal; both are heated until the glass wets the metal, flows or spreads out to form a smooth glossy surface, and forms a chemical bond at the interface, at which time the composite is removed from the heat and allowed to cool. That is about all anyone can say about the process. The rest is about procedures and techniques: how to prepare the metal and the enamel, how to apply the layer of enamel, etc. There are various methods for these procedures.

Enameling begins with cleaning a piece of metal such as copper or fine silver, which are the easiest to enamel. You may wash or screen an enamel to remove the fines. The base coats of enamel are the first layer of enamel on the front and on the back of a piece. Enamels may be applied by sifting, wet packing, spraying or with the thumb and index finger. Spraying is

covered by other enamelists elsewhere in this book. You apply a base coat of enamel to the metal piece, fire it, and let it cool. Then you repeat applying and firing additional coats of enamel, which may be from three to thirty or more. The number of layers depends on the design and the technique. The finished enamel piece is either framed or the exposed metal is polished.

Before you purchase equipment and supplies for enameling, you need to decide: what you will make, what will be the largest size piece, what technique you will use, and how many duplicates you want to make. For the beginner, about an 8" electric kiln on a 110V line is probably the most versatile. The size of the kiln determines how large each enamel piece can be. It is best to have at least a one inch margin for the largest piece inside the kiln because the heat is hotter nearer the wires. You can also buy a hot plate kiln or make one with a solid surface hot plate and a small Pyrex pot with a handle to use upside down as a cover. The hot plate kiln used to be a Trinket Kiln. A setup can be made for a torch. Kenneth Bates, in *Enameling: Principles & Practice* (1951), shows a kiln made from metal coffee cans. Deborah Lozier and Edward J. Friedman explain in this book how they use a torch.

The equipment you purchase, based on the decisions you make, determines the workshop space you need. The ideal workshop has a separate area for each function: grinding and buffing metal, sink washing with a work table, firing, enameling, hand finishing and packaging. I am describing my workshop because it is more than adequate. Charles Jeffery said he made his cloisonné jewelry with a kiln on a desk in his bedroom.

When I began enameling, I used our small pantry, which had a sink. After three years with my 110V 8" kiln that had a pyrometer, I graduated to the basement and eventually to my 16" kiln on a 220V line with a door that now opens from right to left. My favorite kiln had a door that opened up. The 16" kiln was re-bricked about ten years ago with 2300 brick. The 2300 brick will withstand a temperature up to 2300°F. The kiln has a pyrometer in the right hand back corner of the top. It is good to have a spare set of wires for the kiln. Kiln dimensions are usually given width x depth x height. If only one dimension is given, it is for a square floor inside the kiln.

The floor of the kiln is protected from enamel drippings with four 6" x 6" bisqued ceramic tiles that have been coated with kiln wash. The kiln wash comes in powder form and is mixed with water to a thick paint. I coat about eight of them with a 2" wide inexpensive brush and let them dry overnight. I purchased a box of 25 and foolishly did not coat them first with the kiln wash. There are now other items to protect a kiln floor. Any enamel on the kiln floor will get soft every time you fire the kiln, and the feet of the planche will stick to it. Many enamelists vacuum the inside of the kiln because the 2300 brick is apt to shed.

The kiln wires move. They expand as they heat up and contract as they cool. My kiln has the exposed wires sunk in four rows of slots around three sides. If the kiln overheats, a part of the wires could expand and move up and out of its slot. You should push the wire back into its slot only while it is hot, but with the kiln turned off, as the wires become brittle over time. I always wear a cotton smock with long sleeves or a long sleeve cotton shirt with an apron in the workshop.

When I receive new wires for one of my kilns, I make certain they are labeled for the proper kiln before I store them. They are coiled, but need to be stretched to fit the kiln. There is a formula for the length of wire needed. The man who would rewire my 16" kiln wrapped a piece of leather around one end of a wire and placed that end in a heavy vise. He made a mark in the floor across the room from the vise and stretched the wire to that mark.

To purchase replacement wires for your kiln, in addition to the make and model number, you need to know volts, amps, inside kiln measurements, the number of elements and the kind of switch or switches. Switches come hi-lo-med and infinite. Most potters I know rewire their own kilns.

If you visited my studio this is what you would see. My 16" kiln was purchased with a metal stand that has a shelf near the bottom. I keep the firing forks and asbestos gloves there. Because the metal casing of my kiln gets hot, it is 8" away from the wall and at right angles to a 6' sideboard that has a fireproof covering. On top of the sideboard is an assortment of squares: a 12" square x 1" marble slab, a ceramic kiln shelf, and two 3/8" thick steel 14" square plates. To the right of the sideboard is another table with old irons and two heavy weights I had made. On the wall above the sideboard are various sized hammocks I use for firing a plate upside down or a flat piece by opposite edges. To the left of the kiln is a table with a large stainless steel stove protector pad on it

Below:
The Firing Area- A #169 Norman kiln with a 16" x 16" x 9" chamber. The door supports broke and could not be replaced, so the door now opens from right to left. Pyrometer is on top right-hand corner to allow for the full use of the top for drying. There are two infinite switches, one for each element.
A worktable, a right angle to the kiln, is the same height as the inside floor of the kiln. The table is covered with a sheet of ¼" asbestos that was cut to measure. It may no longer be sold. There are two squares of marble, a square of Carborundum, a 14" x 14" x 3/8" steel plate, and the same size steel weight with a handle standing against the wall.
The firing fork has a domed hand protector. The plate on the hammock was fired backside up with the counter enamel on it. The wooden handled stiff spatula is used as a left hand when moving hot pieces with the fork in the right hand.
A blacksmith made the weight. The additional wooden handle was added to eliminate the need for wearing an asbestos glove, because the weight absorbs heat from the metal. The weight is used to correct any warping. The plate in the foreground has just had a first firing of enamel. The three-pronged stilt, which supported it on a planche in the firing, is removed easily when the piece is cool.
The white wire mesh hanging on the wall is a planche that was coated with kiln wash to cover the drips of enamel that could not be removed. *Photo by Bill Byers*

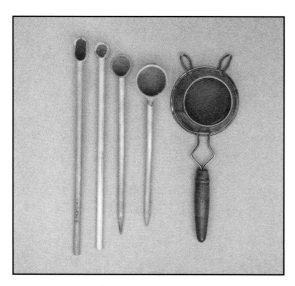

Sifters- From left to right: the oval, ¾", and 1" sifters have 80 mesh screen. The ½" sifter has 40 mesh screen, and the old-fashioned 2" strainer has 60 mesh screen. I made sifters from plastic, pint-sized measuring cups with handles; just had to cut out the bottom of the cup before melting it onto the 60 mesh screen.
Photo by J.A. Perry

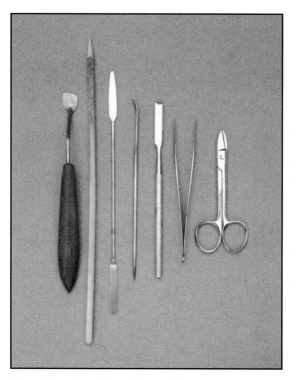

Enameling Tools- From left to right: painter's spatula with broken tip, sharpened chopstick, three dental tools (firest with flexible ends, second with a narrow scoop and a point, third a stiff spatula), tweezers slightly filed to dull the tips, and straight bezel shears.
Photo by J.A. Perry

for pieces ready to be fired. I usually enamel in a production style for the base coats, which I will explain in my section on overglaze direct painting.

For firing equipment, I have 6" square nichrome mesh planches with the four corners bent down and two 6" x 12" ones with two sides bent down. The planches are now available in stainless steel mesh. My firing forks have a guard for the hand so I do not need to wear a glove when firing. For firing a piece right side up, I use a three-pronged Atlas ceramic stilt or a wire one. When I could not find any more of the ceramic ones, I bought all sizes in four-pronged metal wire stilts. I cut off one prong and bent the stilt to form a triangle. The three points balance better than four points. The other tool is a wooden-handled spatula with about a 10" stiff blade. The spatula is my left hand when I remove the hot enamel from the planche. There is one beside each of the two-planche setups for production firing. The double set tub, with a mixing faucet, has a salad cutting board suspended across one tub. The board is convenient for stoning cloisonné pieces under slowly running water and also for scrubbing six switch plates with Penny-Brite copper cleaner. To the right of the sink is a table with a dish strainer and across from the sink is a 5' table for brushing black crackle (now called liquid form enamel) as a base coat on the backs. Spaced on the table are glass jars to support the pieces when the crackle is applied to the backs and a container for the large brush, a sharpened chopstick, and an iced tea spoon. The brush is to apply the crackle, the spoon is to stir the crackle, and the chopstick to sgraffito my name in the center back of each plate or bowl after the crackle coat dries.

The grinding and polishing equipment are connected to a large dust collector that I bought from a dental supply company. I have two belt sanders: one takes a 1" x 36" fine emery cloth belt and the other a 6" x 48" belt. I use the wider belt sander to bevel the front edge of plates. To use with buffs for cleaning and polishing, I have one motor with two regular-sized spindles and one motor with a 10" long spindle that a friend made for me to simplify cleaning the inside of bowls. I have another homemade setup of an electric drill that holds a 5" emery fine disk for sanding the back edge of plates. My friend used an old washing machine motor for that one.

For cutting, there is a Beverly shears and a large vise that will hold a pair of airplane shears. I also have my jeweler's bench with the typical equipment for making jewelry and soldering. A 6' table for sifting or wet packing enamels holds sifters, bruches, pens, bottles of enamels for base coats and a supply of white typing paper. I use white paper under the piece when I am sifting so I can see that the enamel has no unwanted specks. A smaller table holds petri dishes with the overglazes for painting. Two closed closets hold the bottled enamel supply.■

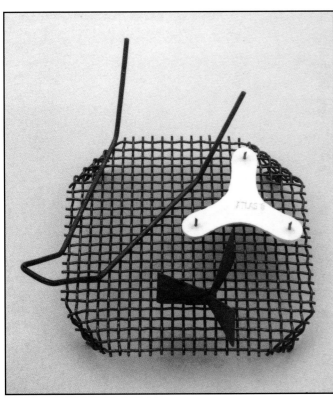

Left:

Planche, Hammock, Ceramic Stilt, and Trivet- These firing supports come in various sizes. The hammock, ceramic stilt, and trivet are shown on a 6"square, nichrome mesh planche that goes in the kiln. The hammock supports a flat piece and an upside-down plate or bowl. The stilt supports a right-side-up piece. The small trivet supports a piece of jewelry on the edges. *Photo by J.A. Perry*

Bottom left:

Kiln Tools- From left to right: a firing fork made with an 8" bowl for a guard and a long rod and long tines; a stiff bladed, wooden-handled spatula that is used in the left hand to help support a hot enameled piece as it is moved; a flat-ended chasing hammer used to help restore form to a warped piece; and a firing fork purchased with a hand guard. *Photo by J.A. Perry*

Bottom right:

Brushes- From left to right: 1" greyhound bristles, #3 sable bristles, Hunt 101 nib pen holder, #1 sable liner, #00 sable, and handmade Lebezon kolinsky watercolor brush. *Photo by J.A. Perry*

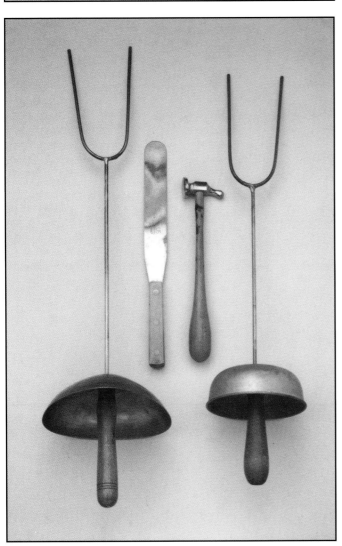

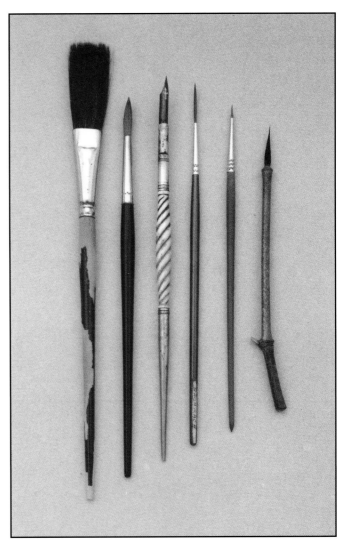

Metals For Enameling

Copper

The most commonly used metals for enameling are pure copper and fine silver. The fine silver is primarily used for jewelry because of its cost, color and the advantage of not producing a firescale coat like copper does when it is fired. The copper is the most used metal for enameling for many reasons: its malleability, its color, availability, and cost. In addition, it usually maintains its shape in the firing process. The appropriate gauge (thickness) of the metal depends on what is being made. Most plates and vessels are usually 18 ga while jewelry can range from 28 ga for a repoussé piece to 14 ga for champlevé. The smaller the gauge number, the thicker the metal. If pure copper is to be ordered from a mill, it is necessary to specify the form, such as sheet, wire, etc., the gauge and the end use (enamel on copper). The *Thompson Enamel Workbook* specifications are "oxygen free, high conductivity copper, conforming to ASTM specifications B-170." I do use the 20 ga sheet copper from a plumbing supply company for pieces up to 4" x 10" with about five firings and I have not had any problems. I have not used it for more firings because the need has not arisen.

Base-coated Steel

I have used steel in flat 12" squares that came with their undercoat and base coat already fired. If I overfired them so that they warped, it sometimes was impossible to flatten them even though they were fired and weighted many times to try to correct the warping. Copper is more forgiving if it becomes warped. Steel needs a special undercoat before the base coat is applied. Purchased with the base coats, the pieces only need to be degreased with a detergent before enameling.

Fine Silver

I usually dome jewelry that I fabricate of 26 ga or 28 ga fine silver or oxygen-free copper. Doming the thin guage metal gives it added strength and keeps the piece of jewelry, especially pins, from being too heavy. I often leave the 18 ga and 20 ga fine silver flat. I do counter enamel all of my enamels. Many enamellists say that there is less probability of the enamel cracking if the base coats have the same enamel.

Cloisonné Wire

I like to purchase both fine silver and fine gold cloisonné wire that is not annealed because it is easier to make a straight line with the wire stiff and it is easy to anneal it for intricate lines. To anneal the wire, loosely and carefully wind it into a three to four inch roll and fire. To test whether the wire is annealed when you take it out of the kiln, bend back one end; if it stays bent, it is annealed and if it does not, it needs additional firing.

Foils

The only foils I have used are the standard 24k gold and fine silver and also a heavier fine silver foil sometimes known as "clutch" silver. Although many enamelists pierce the foil, I never do, even though I have used the full 3" square sheets on enameled pieces. Jean Jenkins places the foil, after she cuts it, on an anti-static sheet. I cut a sheet of foil between two pieces of tracing paper with various sized sharp scissors. I have a 10" pair that I use for long diagonal cuts. The foil is placed on the enameled piece with a water-dampened #1 liner brush. Additional water is added to the piece if the foil needs to be repositioned. The water is drawn off with a paper towel by pressing down on the towel. If the foil does not adhere to the enamel surface after the water dries,

Materials for cleaning copper with a buffing wheel and Lea Compound C- Shown up and around from left: new 3" goblet buff, 5" goblet buff coated with Lea Compound C, open tube of compound stored in a metal frozen juice can, new tube of Lea Compound C, glass jar of compound paste, table knife to apply paste to buff, and new 6" x 1" cotton muslin buff. *Photo by Bill Byers*

then I add a 1:2 diluted gum and wait until it dries to fire the piece. For the first firing of the foil, I often sift about a 1/2" circle of soft flux on an area that has no foil. This spot of flux will tell me when the foil is fused. If you overfire the foil in the first firing, you will destroy it. After the first firing, I smooth the foil with a glass brush. The foil can be fired higher after there are two layers of transparents over it.

CLEANING COPPER BY BUFFING WHEEL

If you use primarily transparent enamels, any marks on the metal surface will show. A bright surface enhances the brilliance of transparent enamels. To achieve the surface on copper, I use the buffing wheel. **Caution:** Although I have an exhaust system, I wear a nose mask, a facemask, a shower cap and cotton or leather gloves. To be safe, you need to know how to use the equipment.

I use a 5" cotton goblet buff charged with Lea Compound C, a greaseless compound for copper in a metal tube. Cut off about 2" of the tube, remove the wrapping, and place the piece of compound in a small glass jar with just enough water to cover the compound. The jar needs to be tall enough for you to stir the compound after it is soft. Left overnight, the compound absorbs the water and is soft enough to be stirred into a smooth paste. I apply a thick coat of the compound paste to the goblet buff mounted on the spindle of the buffing wheel. A stiff, small, metal spatula works well for spreading the compound on the buff. I usually let it dry overnight to a hard, crusty state. I soften the cutting action of the buff by first giving a few swipes with a piece of scrap copper across the spinning charged buff. You can clean over twelve 6" pieces before the buff needs to be recharged.

You can also apply the Lea Compound C directly to a buff on a spinning spindle at 1725 rpm by holding the open end of the compound tube again the buff. This thin coating on the buff will dry within five minutes and then you can buff the piece. Charged this way, the buff will need to be recharged with the compound after a few pieces.

I buff the front of the piece first and the back second. By buffing the front first, the finger marks on the back will be removed when I hold the piece by the edge to buff the back of the piece. If it is difficult to hold the piece by its edge, I wear clean white cotton gloves to buff the second side.

The Lea Compound C is stored with the cut end standing tightly inside a small, metal, frozen juice can with a wet paper towel in the bottom to prevent the compound from drying out.

CLEANING COPPER IN AN ACID BATH

The first rule for using acids is **add acid to water, not water to acid**. The most commonly used acids for removing firescale from copper, sterling, and gold alloys are commercial grade nitric acid and Sparex 2, an acid-type cleaning compound in granular form. Acid diluted in water is called a "pickle." Jewelers often say, "pickle it." The solution can be used in a hard rubber photography tray, a Pyrex container, or an electric slow cooker. The pickle works best when warm, but it should not be allowed to boil.

Although Sparex2 is considered a "safe" acid, both the nitric acid solution and the Sparex2 will eat holes in fabric. The weaker the solution, the slower the biting action. The recommended solution for the Sparex2 is 10 ounces by weight in warm water to make a quart of solution. I have only used the Sparex2 in the electric slow cooker after soldering silver or gold jewelry. I do not clean the copper with acid, but other enamelists explain how they use it and dispose of the acid.

The other rule when using an acid bath is **use wooden or copper tongs. Tongs made of iron will contaminate the acid bath**. If a deep container is used for the acid bath, wear special, long, heavy rubber gloves when you reach into the tank.

Charging a new 5" x 1" muslin buff with Lea Compound C paste- The spindle was custom-made with extra long length to accommodate buffing the inside of deep bowls with a goblet buff. *Photo by Bill Byers*

CLEANING COPPER BY HAND

Each enamelist has a preferred product for cleaning copper. Among the ones I have used are a liquid dish detergent, a rag soaked in vinegar and dipped in salt, and a scouring powder with a scouring pad. I prefer Penny-Brite. I have used it for over twenty years. It is a copper cleaner with the right ph. I place a number of small pieces, or one 12" plate, on the wooden board across the top of a set tub. This board for cutting vegetables at the sink, is sold in kitchen supply stores. Sometimes I wear rubber gloves. I rub one side of all the pieces with a scouring pad and Penny-Brite, then turn them over, clean the other side, and rinse them well. The piece is grease-free if the water sheets off (does not bead). I place each piece in the dish strainer, dry them with a cotton towel, and if I am not coating the backs with liquid black enamel at the time, I wrap each piece in paper. If I am ready to brush on the crackle, I place each piece, back side up, on a bottle on the nearby table *(see photo on page 22)*. If the water, gum, or crackle pulls apart (coagulates) while you brush it on, you can clean the spot with a little enamel, crackle, or saliva to remove the grease from that small area.

CLEANING FINE SILVER

I clean fine silver in the kiln. I first put a number of fine silver, jewelry size pieces, on trivets into a 1500°F kiln to remove any grease or discoloration. I peek into the kiln to check whether they are clean. Fine silver is silvery white when clean. When the pieces are cool, I rub one piece at a time with 000 steel wool. A piece is placed on a small sheet of typing paper, and one corner of the paper is turned over a small part of the piece to keep the oil from my finger off the metal. My finger keeps the piece from moving. I rub in one direction over the face of the piece, usually from top to bottom. The clean pieces are placed face side down on a clean sheet of paper about 2" apart for a base coat of very thin crackle. I use the fine silver for cloisonné pins and pendants.■

KARAT GOLD ALLOYS

STUART GRICE, *Mill Products Director, Hoover & Strong Inc.*

Stuart Grice is the Mill Products Director at Hoover & Strong Inc, VA. Previous positions: Metallurgical Manager, Cookson Precious Metals Ltd., Birmingham, England. Senior Metallurgist, Johnson Matthey Jewellery Ltd., Birmingham, England. Higher National Diploma in Applied Physics (HND). Materials Science (BSc). Representative, British Jewellers Association, on the International Standards Technical Committee ST1/53, 1997-2000.

Karat gold alloys fall into several different categories or "marks." In the United States these are 10kt, 14kt, 18kt and 22kt, with Europe having 8kt and 9kt, the Middle East 21kt and the Far East 24kt. A "karat" gold is not pure gold, but an alloy that contains 24th fractions of gold by weight.

10kt	10/24 gold	41.67% gold
14kt	14/24 gold	58.33% gold
18kt	18/24 gold	75.00% gold
22kt	22/24 gold	91.67% gold

Karat gold alloys are commercially available in numerous colors and hues. The most common colors are yellow and white, with pink, red and green becoming popular with fashion trends. The majority of higher karat yellow, pink, red, and green alloys are comprised of gold, silver, and copper. The ratios of each metal can be varied to achieve the different colors. Alloys in the lower karats also contain zinc, generally in quantities up to 10%. Zinc is added for various reasons, but in the lower karats it is a color enhancer and gives yellow alloys a warmer feel. White alloys have are two major sub-families to be considered: gold/nickel and gold/palladium. The palladium alloys do not require the presence of zinc; however, this does not mean zinc will not be present. The nickel alloys usually include zinc up to about 6%. The melting points of typical karat gold alloys are as follows:

10kt	Yellow	1500°F
10kt	White	1770°F
10kt	Red	1715°F
10kt	Green	1530°F
14kt	Yellow	1500°F
14kt	White	1700°F
14kt	Red	1700°F
14kt	Green	1600°F
18kt	Yellow	1580°F
18kt	White	1650°F
18kt	Red	1630°F
18kt	Green	1780°F
22kt	Yellow	1830°F

Enameling fine gold has an obvious advantage: gold does not oxidize under normal circumstances. When considering the alloys of gold, oxidation of base metal additions in the alloys (copper, zinc, nickel, cadmium, tin and indium) makes life somewhat harder. The piece must be heated to bring the base metals to the surface and to oxidize them. Then follow several warm picklings, acid baths usually about 75°F. Most jewelers put their pickle in a small electric crockpot.

The first pickle, usually sulfuric, removes the oxides of base metals; i.e., they revert from oxides to base metals. The second pickle, nitric, removes the base metals themselves, but also silver, which is not a base metal. These metals migrate to the surface, but they can only travel a certain distance, and oxidation will only occur to a certain depth below the surface. Therefore, the process must be repeated several times. After the last nitric solution pickle, a fine gold layer is on the surface. Not all the base metals have been removed; the majority remain below the surface.

Copper oxide prevents enamel adhesion and also results in color problems, but by far the most inconvenient addition for enameling is zinc. Not only does zinc oxide prevent adhesion and lead to color problems, but also it will be released from the metal surface as a vapor at enameling temperatures. These problems will occur even if the surface has been pre-cleaned and lead to unavoidable problems once the enamel layer is in place.

Consequently, the best karat gold alloys for enameling are either zinc-free or contain very low zinc concentrations, typically 2% or less. As a metallurgical purist, I say 1% zinc. Alloys containing higher levels can be enameled, but more care is necessary in the preparation stages. All work pieces must be depletion gilded, sometimes known as fire gilding, before the first layer of enamel is applied. In this process, the work piece is annealed without a fire coat protection to allow the base metal — usually copper, zinc and sometimes nickel if white gold alloys are used — to oxidize. Sometimes quenching the work piece into water once the red glow has disappeared helps; but be careful of this step if using a 14kt or 18kt nickel white alloy since cracking can occur when quenching. The oxides are then removed in an acid pickle bath of 10% sulfuric, followed by base metal removal by further pickling in 10% nitric acid. A caution: if you leave the work piece in too long, all the base metal will be removed and the work piece will disintegrate. The various pickle baths will leave a fine gold enriched surface. The pickling process must be repeated several times to ensure all the base metals are removed from the surface. As a general rule, the lower the karat, the more repetitions since a higher content of base metals is present.

It is wise to note at this point that problems can occur when enameling high karat red gold alloys. With 10kt red alloys only the oxide issue requires consideration, while for 14kt, and in particular 18kt red gold alloys, things can be a little different. Internal strains occur on cooling from annealing temperatures because of atomic movement. These factors make the metals hard and can also distort the work piece, and in extreme cases cracking can occur. Quenching from above 770°F will reduce these possibilities, but not once the first coat of enamel has been applied. 18kt red alloys are probably the most difficult to enamel successfully.

As a general rule, when choosing a karat gold alloy for enameling, the higher the karat the better, since as the karat level goes down, the base metal content goes up, particularly for zinc. Yellow alloys are usually easier to enamel than whites and reds. All karat gold can be enameled with transparent enamels. If sufficient depletion gilding has been done, there is no need to fire a layer of opaque enamel; fuse fine gold over this opaque layer and then transparent enamel to finish. The fine gold layer resulting from the depletion gilding means that transparent enamels can be fired directly onto the work piece. The less the base metal additions in the alloy, the easier the process will be and the less potential for failure there is. Most 18kt yellow alloys have little or no zinc present and are therefore the best alloys to use. Problems can occur when using opaque red, green and yellow enamels, which tend to discolor to brown if overheated.

When ordering karat gold alloys for enameling, ask about zinc content and what is recommended as the best alloy in the karat range you can afford. Most metal suppliers do not carry specialized enameling karat gold alloys, so cost should be no different from standard alloys. At Hoover & Strong, the minimum weight order for sheet is 2 dwts for gold, in any karat or color. As a guide, a 1" square sheet in 20 ga will weigh approximately 5 dwts in 18kt yellow, 4.4 dwts in 14 kt yellow and 3.9 dwts in 10 kt yellow. For 22 ga sheet, these weights become 4 dwts, 3.5 dwts and 3 wts respectively.

Enameling techniques and considerations required for base metals also apply when using karat gold alloys.

 • Make sure all soldering operations have been completed with a suitable grade solder, i.e., with a melt point greater than the enameling temperature. Always keep solder joints to a minimum since karat gold solders often contain zinc and also cadmium, which behaves like zinc when heated and will have similar associated problems.

 • It is usual to heat treat the work piece prior to enameling in order to relieve any residual stress present. The preheating will prevent distortion that may otherwise occur during the enameling firing cycle. This can be done at a low temperature, around 660°F, for a minute or two before the depletion gilding process. Most enamellers will depletion gildthree to four times to ensure no base metals remain in the surface. This will also totally stress relieve the work piece and ensure that no distortion occurs. The actual number of times required will become an "experience" thing for each enameller, who will become comfortable with a particular suppliers product. Four depletion gilds should be the maximum required under most circumstances. Better too many than too few.

• After the final depletion gild, the work piece must be meticulously cleaned. Traditionally this takes place by pickling in warm 10% nitric acid to remove any traces of oxides, grease and dirt. Some enamellers prefer to use 10% sulfuric acid since it is less reactive and will produce a suitably clean surface. Nitric acid will result in a superior surface, if required.

• Either method requires neutralizing the piece in an alkali bath such as 10% sodium bicarbonate to remove any traces of pickling acid. Ideally do this in an ultrasonic bath. Any entrapped acids remaining in surface porosity are areas where liquids may become entrapped by capillary action that will eventually result in corrosion products and will lead to discoloration.

• An alternative cleaning method, which is increasing in popularity, is electrolytic stripping. This provides an excellent surface finish for enameling, but equipment costs have to be considered.

Typical problems when enameling karat gold alloys:
• Poor enamel adhesion due to an inadequately cleaned surface.
• Poor enamel adhesion due to the presence of oxides at the surface of the work piece. The oxides are usually from copper, zinc, or nickel.
• Enamel discoloration as a result of oxides at the surface of the work piece.
• Enamel discoloration resulting from entrapped pickling acids in either surface porosity or areas where residues may be retained by capillary action. The entrapped acids will result in the generation of corrosion products, which can be seen through transparent enamel. If a work piece or component has been investment cast, the probability of porosity at the surface is greatly increased, for shrinkage and gas porosity are known defects from this manufacturing method.
• Work piece failure due to assembly using a solder of insufficiently high melt point.
• Partial liquidation of the work piece due to the melt point of the alloy being exceeded in the enameling process.

In summary, when choosing gold alloys for enameling, use a zinc-free or low zinc alloy if possible. Be aware of the increased potential for copper oxidation and distortion, particularly in 18kt, from red alloys. It is not intended to suggest that these alloys cannot be enameled, only that they need to be treated with respect and great care.■

S.Grice, "But I've Always Done It This Way. Technical Support It Makes a Difference." 12th Santa Fe Symposium on Jewelry Manufacturing Technology, 1998, Met-Chem Research Inc.

"Jewellery Enamels and Their Applications to Copper Based Metals." Johnson Matthey Blythe Colors Data Sheet C17.

W. S. Rapson and T. Groenewald, *Gold Usage,* Academic Press, 1978.

THE MANUFACTURE OF ENAMELS

WOODROW CARPENTER

Woodrow Carpenter, owner of Thompson Enamel, Inc., fired his first enamel in 1935. In 1950, he began manufacturing and selling enamel colors. In 1981, he purchased the Thomas C. Thompson Company. In 1982, he began publishing Glass on Metal, *published by Thompson Enamel, Inc. He founded the Society in 1986.* Glass on Metal *has many technical articles on the composition and properties of enamels and their firing written by Mr. Carpenter.*

Woodrow Carpenter

Minerals such as silica, soda ash, potassium nitrate, borax, calcium carbonate, etc. are weighed according to each enamel formula, well mixed, and loaded into a preheated fire clay crucible. The batch is heated to a selected temperature for a sufficient length of time to melt the minerals, forming a viscous liquid, and continued until all gases released during the decomposition of the minerals are eliminated. The enamel is removed by ladling or pouring it onto a thick iron plate where it cools to form "cake" or "lump," into water where it is shattered to be called "frit," or through water cooled iron rolls to be called "roll quenched." The cake and roll quenched are also known as air-cooled. The minerals mentioned above will produce colorless transparent enamels. Changing the proportions of these minerals will provide a wide range of firing temperatures and thermal expansions, as well as small changes in gloss, surface tension, and other properties less well known.

Other minerals may be included in the colorless transparent enamel formula. Some crystalline minerals have a low solubility in glass. These insoluble crystals will decrease the transparency. Depending on their size, number, and index of refraction, the decrease can range from a slight cloudiness to a dense opaque white. This range of opacity has been described in terms such as translucent, opal, opalescent, and opaque.

Several manmade ceramic pigments are added to control color. When added in the absence of insoluble crystals, the result is a transparent color, frequently called translucent. When insoluble crystals are present, the result is an opaque color.

The composition of lead and lead-free enamels differ only in that the former may contain as much as fifty percent lead while the latter contains no lead. Any change in an enamel's composition may require some change in the procedures used by the artist who wishes to obtain the same end results. Different enamel compositions have different optimum firing ranges to produce their best visual properties such as gloss, clarity, and smoothness. This firing range is rather narrow, usually a spread of 30-50°F, and a length of time at that temperature, which must be determined by the artist.

Furnaces that are slow coming back to temperature after inserting the piece may require a total of five minutes or more to provide the proper length of time at the optimum temperature. Artists who fire by observing the surface of the piece, with little regard to what their indicating pyrometer reads, have had no trouble firing lead-free enamels. Artists who arbitrarily fire at what they think their indicating pyrometer should read and at a predetermined time have problems switching from one enamel to another even if both are lead enamels.

I cannot tell you what your pyrometer should read or how long you should fire to obtain a clear flux coating of lead-free enamel on copper. I can tell you that in my furnace with my controlling pyrometer set at 1500°F the length of firing is four minutes. If you can arrive at a comparable fire, the result will be equal to or better than any lead-bearing enamel. When using opaque lead-free, I set my controlling pyrometer at 1500°F and fire for three minutes. Using a controlling pyrometer eliminates the need to look at the enamel while in the furnace once you determine the optimum fire (temperature and time).

We know many artists who feel their lead-free transparent colors are superior to lead-bearing enamels. On the other hand, many feel lead-bearing opaques have a little more gloss, but they continue to use unleaded for health reasons.

Lead enamels can be fired as a subsequent coat over lead-free enamels; lead-free enamels can be fired as a subsequent coat over most lead enamels. They can be sandwiched. For example: a lead flux, lead-free color, followed by a lead color. Again, the secret is in the firing. All of this was done quite extensively in the sixteenth century in Limoges. Prior to that, the eleventh century champlevé enamels by the Limoges and Mosan masters were lead-bearing white, yellow, and green, while all other colors were lead-free.

There is one caution. A complete coat of lead enamel can be applied over a fired coat of lead-free; however, if only one or two grains of lead enamel fall onto a lead-free surface and fired, a pit will result. Thompson stopped making lead enamels because of OSHA's workplace restrictions on the airborne lead.■

THE ENAMEL MATERIAL AND ITS APPLICATION

Enamel comes in a number of forms: lump, string, liquid, and powder, as well as in the optical qualities of transparent, opaque, and opalescent. The important factor in selecting an enamel is that it be made for the metal you are using. Enamel expands as it is fired and then contracts as it cools. This is called thermal expansion. The metal on which the enamel is fired must expand and contract at a slightly higher rate.

Enamels are sold in assorted lump forms and in meshes, probably as coarse as 10 mesh and as fine as #325. Some enamelists use the fines for a painting technique. I principally use 80 mesh powder, overglazes, and the 20 mesh in transparents for some jewelry.

Enamels are manufactured in soft, medium, and hard fusing, which refers to how they fire. The soft enamels fire the most quickly. Some enamelists refer to the soft enamels as delicate. In Thompson's catalog, most of the 80 mesh enamels for copper, steel, silver, and gold are listed as medium fusing. Only the flux and the black have a listing as soft; the flux has an additional listing for hard.

When I first studied enameling, I was taught to use only the 80 mesh soft flux as the base coat and then the medium fusing enamels for subsequent coats. It was not until I concentrated on painting with overglazes that I changed to using medium fusing enamel as a base coat because it did not bubble up through subsequent layers of enamel. It is now accepted that you first apply the hard firing enamels, then medium over those, and the soft enamels for the top coats.

Some enamelists do not remove the fines from their enamels. To remove the fines from ground enamels, i.e., clean them, you either wash them or screen them through a stack of various mesh screens. To use the screens, you stack them with the coarsest mesh screen on the top and place a penny in each one to help move the enamel. You shake them until the fines are at the bottom. Then you put each screened enamel in a separate labeled container.

To wash the enamels, use either your tap water, depending on its quality, or distilled water. I wash only the enamels I use for jewelry. Place some enamel in a jar, add water, stir, let it start to settle, and pour off the milky substance in the top of the jar. This process is repeated until the water in the jar with the enamel is clear. I seldom wash more than ¼ cup of an enamel at a time and often just a teaspoonful. I spread out the washed enamel on two stacked sheets of clean paper near the back of the kiln top and cover it with another sheet of white paper to keep it clean while it is drying.

An early instructor of mine had a wide-mouth gallon bottle into which we poured the milky substance left after we washed the enamels. When enough waste was accumulated, the water at the top was poured off, some fresh water added, stirred, let settle, and the milky water poured off. When dry, this discarded enamel was added to the counter enamel.

Enamels are best stored in bottles with a screw-on lid, labeled with their number, manufacturer and mesh size. If the enamel has been washed, I add that to the label. The labeled bottles, arranged by color and transparent or opaque, are kept in a closed cabinet.

COLOR SAMPLE BOARD

Many enamelists have enamel color sample boards, one for opaques and one for transparents. The opaques only need two siftings of the colors you own. The transparents, often on 1/2" x 3" of 20 ga or 18 ga copper, show the transparent color on the bare copper, over flux, medium white, silver foil, and gold foil. You have to divide each piece into five sections: the top fifth is for the transparent on the copper. You can either coat that section with Scalex or leave it uncoated to clean after the first firing. The next section down is 80 mesh medium flux, the third one down is medium fusing white, and the last two sections are for the silver and gold foil. The sections under the foil can be the flux or the white enamel. Fire for the base coats and then fire a transparent enamel color over the entire piece.

For fine silver samples, I clean 1/2" discs in the kiln, brush a thin coat of crackle on the back, and sgraffito an enamel number in the dry crackle. I sift and fire a covering coat of soft flux. Then I sift and fire a covering coat of transparent enamel color. I keep the fine silver samples in a plastic box.

Jean Foster Jenkins' sample board.

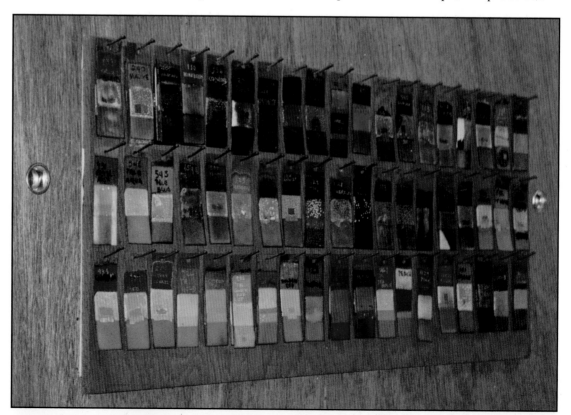

BINDERS (Enameling Gum)

Although enamel can be applied dry, there are many times when you need to mix the enamel with a binder that is an enameling gum adhesive. Years ago, the only adhesives that enamelists were taught to use were gum tragacanth and agar. I remember buying tragacanth flakes, dissolving them in water, and then storing the solution in the refrigerator. When Klyr-Fyre came on the market, I bought a gallon of it because it had an unlimited shelf life. Other enamelists do use CMC (carbo-metho-cellulose). Jean Jenkins gives a recipe for five gallons.

APPLYING ENAMELS

Enamels may be applied by sifting, wet packing, spraying or with the thumb and index finger. I do not own a sprayer, but its use is covered in this book by other enamelists.

MAKING A SIFTER

Except for an old fashioned, metal-sided, 2" diameter, flat-screened bottom tea strainer with a wooden handle, I make my own. Other enamelists do, too. My sifters were adapted from the one Kalman Kubinyi made. I cut about a 1" piece of plastic tubing, diameters ranging from 1/4" to 1-1/2", and shape one end of a 1/4" wooden dowel with a half-round file to fit the tubing. I place a piece of screening, larger than the tube, against a hot electric iron and then press the plastic tube's open end against the screen. The two are held together until I see the plastic melt into the screening. I have an old iron that I keep for this purpose. (Sarah Perkins says that she uses a spatula for fusing the screen and the plastic.) The unit is removed, the screen is trimmed close to the plastic, and then the shaped end of the dowel is attached with Duco cement to the tube. The tube is placed on the workbench with the tube flat on its side and the dowel handle upright and supported against a set of drawers for the cement to dry overnight.

SIFTING

I sift over a stack of two or three sheets of white paper. I use clean white paper to see any speck in the enamel to be removed to keep my enamels clean. The sifter, half filled with an enamel, is on the paper. Students learn to sift an even coat by covering most of the sheet of paper a number of times with counter enamel in a 2" sifter. I sift by tapping on the handle of the sifter with my index finger. If you hold the sifter close to the paper, you limit the spread of the enamel as it falls from the sifter. If you want an overall fine layer, then you hold the sifter up higher and tap it lightly. You have learned to sift the enamel when you can do it without thinking about it.

When you need to wet the metal before sifting on the enamel, you can use an airbrush, a soft brush, or a hand spray bottle. I remember Kenneth Bates telling students in a workshop that you aimed the sprayer at the ceiling and let the solution fall on the piece as the gentle rain from heaven. If you are using a hand spray bottle, set the nozzle to the finest spray and pump it a few times before spraying the metal. If you do not have a spray booth and you are doing a lot of spraying with a gum solution, it is advisable to cover a section of the floor with newspapers. A floor wet with gum solution is very slippery. I do most of my spraying with water, so there is no problem with a slippery cement floor.

WET PACKING (Inlaying) 80 or 100 mesh ENAMEL

I pick up damp enamel with the tip of a #1, #2, or #3 sable liner brush and place it in the top left hand area of the design. With the side of the brush, I level the enamel to about 1 mm thick, wipe off the brush, and again with the side of the brush, draw off any excess water. The next damp color is placed almost up to the first color and then pushed against

Applying black crackle (liquid form enamel)- Black crackle is being applied to the back of a cleaned copper plate. The brush is a 1" greyhound. Although half the length of the well-charged brush is held against the pate, no pressure is being exerted. The liquid enamel is feathered onto the plate. *Photo by Bill Byers*

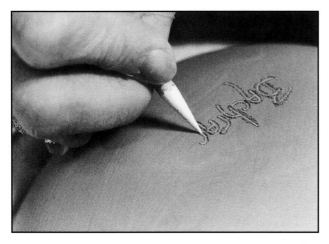

Signature- A signature is being sgraffitoed with a sharpened chopstick in the dried black crackle on the back of the plate. The plate will be tapped on its edge to remove the loose, dry crackle. *Photo by Bill Byers*

the first color, evened out, and so it continues. The size of the bush you use depends on the area an enamel color is to cover. You might need to add a very little bit of diluted enameling gum if you are working on a sloping surface. If you are going to wet pack a piece that is larger than 4", you should start in the far corner and progress diagonally down to the opposite corner. That way your hand does not disturb the enamel while you are applying it with the brush. Other enamellists use other tools for wet packing. Just try them all.

INDIAN SAND PAINTING (THUMB AND FINGER APPLICATION)

This technique takes practice. The space between the first and second joint of your index finger is placed next to some dry 80 mesh enamel on a sheet of paper. (I have not used it with other enamel meshes.) Your thumb slides across the enamel and holds it against the flat area of your index finger. You start releasing the enamel by sliding your thumb forward. You begin either at the top or the bottom of the line you are making.

APPLYING LIQUID FORM ENAMEL (CRACKLE) AS COUNTER BASE COAT

Although the name of the material has been changed, what I still have and use was called "crackle." The name refers to what happens once this liquid form enamel is applied and fired over a previously fired coat of 80 mesh soft enamel. The liquid form breaks up or cracks in the firing because the soft enamel underneath expands first. I suspect it was a porcelain slip made for pottery. I still call it crackle.

When I last purchased Thompson #772 Black it came in a plastic gallon jug. I poured the liquid into a big basin and cut off the top of the jug to spoon out the mass at the bottom of the jug. After a lot of stirring with a long cooking spoon, I was able to get the material into solution. I divided the crackle into many wide-necked, glass pint jars with screw-on covers, stirring the crackle before each ladle-full was poured into a jar. The jars were filled to within 1" of the top to leave room for stirring the crackle before using it.

After my course with Doris Hall in 1955, I used the black crackle as the base coat on the back of all my plates, plaques, and shallow bowls. Originally, I held the piece over a bowl and used a small ladle to pour the crackle on the piece. Sometimes I would move the crackle with the convex side of the ladle.

I soon found that it was simpler to use a well-charged, large brush for applying the crackle. I still apply the crackle with a 1" greyhound brush. When you first receive the liquid form enamel, stir it well with a slender spoon and test a brush full. If it is too thick, gradually add a very small amount of water while stirring. Experience will tell you the right consistency. The brushing consistency is a little thicker than the one for pouring. It used to come in a binder, so if I thinned it too much, I would set it aside to evaporate. Now if that happens, I add a little Klyr-Fyre, stir it well, and it thickens. Sometimes I even have to add a little more water. For my undercoat of counter, the crackle does not need to be an absolutely smooth coat because I will cover it with a counter enamel that is mainly opaque enamels. If I need that smooth a coat, I probably would have bought an airbrush and set up a spray booth.

The piece to be coated with crackle is placed on a jar so it is at least 2" above the table; the side to be coated, usually the back, is facing up. The table is covered with whit paper. If I am coating only one piece and the crackle is very thick, I just wet the brush and swish it around the top of the crackle in its jar and then brush it lightly on the cleaned metal piece. When the crackle is dry, I sgraffito my name with a sharp chopstick in the center back of the plate. If the dried crackle powders off on your hand as you hold the piece to sift enamel on the front, it means it does not have a sufficient adhesive. You can either bear with it, or remove the dried crackle and recoat the back of the piece. The crackle needs to have about a half teaspoon of Klyr-Fyre stirred into the jar. When applying the crackle with a brush, you pass the brush over the piece as though it were a feather. If you bear down on the brush as you stroke the metal, the crackle will not cover smoothly.

When applying the crackle to a round plate, the last swipes with the brush are around the edge of the plate. When I have finished using the crackle, I wipe around the top of the jar, add a little water, and screw on the lid. My old jars have metal lids so I put a piece of white paper over the top before I screw on the lid. The brush and the spoon get a good washing. The brush is used only for the black crackle.

THE FIRING PROCESS

If my electric kiln had a rheostat, I would set it at 1500°F and be able to forget it until I was ready to fire. I should have installed a rheostat; I do know why I did not get to it. I now turn on both infinite controls to the highest number and set a timer for 25 minutes as a reminder to check the kiln temperature. If I should forget, and I haave, the inside of the kiln is almost white and the wires may be burned out. When the temperature is a 1500°F, I turn the controls down to hold at about 1350°F. The longer the kiln is on, the more heat the bricks absorb and although the pyrometer reads the same 1500°F as it did when it first reached that temperature, it recovers faster. A hear-saturated kiln is preferable.

Every time you open the kiln door, the temperature drops. If the kiln furniture is cold when you put the piece in the kiln, the temperature will drop even more. Although 1500°F is 1500°F, I think of a kiln that just came to temperature as a cooler kiln than one that has been at temperature for over an hour. The inside back of the kiln is hotter than the front part, so I often rotate a large piece for the firings. The firing is hotter closer to the wires. I am careful to limit size in order to leave one inch from the wires.

The recovery time of the kiln's temperature is important because when you place an enameled piece in the kiln, the previously fired layer of enamel cracks and mends itself. If the kiln is too cool, the cracks remain.

SIFTING 80 MESH ENAMEL AND FIRING FRONT AND BACK BASE COATS

With this method, the front and the back of the enamel is fired in the first firing. If the piece is flat or shallow, I spray water to hold the sifted enamel in place. If the piece is a deep bowl, I spray with diluted 1:3 Klyr-Fyre. Holding the underside of the piece with the dried crackle resting on my fingers, I spray lightly, sift, spray, sift, and spray, applying a damp enamel coat that covers the copper. I use the 2" 60 mesh sifters for the base coats. Three of the 2" 60 mesh sifters are labeled to be used only for flux, white, or for counter. If I have used only water and sprayed lightly, the piece can be fired almost immediately. If I have used enameling gum, the piece has to be dry. If you are in a hurry, you can set the piece, right side up on hot kiln furniture. If the gum is not dry, you will see some steam rise up when you place it in the kiln. You can also put the set-up piece in the hot kiln for a second, take it out to check for steam, and then repeat until there is no steam. I fire at 1500°F until the front base coat is smooth. When I remove the piece from the kiln, I check that the crackle on the back is shiny and smooth; if it is not, I quickly put it back in the kiln for an additional firing. If the crackle is not fired adequately, it will chip off as the piece cools.

The second firing completes the counter enamel of the back. For my counter, I mix about 1/3 Thompson #124A hard black 80mesh leaded enamel with 2/3 left over mixed enamels. If one enamel color contami-

Sifting base coat on front- The first sifted layer of enamel on the bare metal, the base coat, is being applied to the front of the plate. The enamel being sifted is 80 mesh flux. The enamel in the sifter covers the bottom of the sifter and fills only half of the sifter. The white enamel in the other sifter, sitting on a sheet of white paper, will be sifted over the flux coat before the piece is fired. The fingers of the supporting hand are under the tray, which has a coat of un-fired, dry black crackle. The plate is held tilted to have the enamel fall on the edge. The index finger taps on the sifter to apply the enamel. The excess enamel on the paper is from sifting beyond the edge of the plate and from setting the sifter down to even-out the enamel in the bottom of the sifter. This sifter is an old-fashioned strainer that has a flat 60 mesh screen on the bottom. *Photo by Bill Byers*

Sifting on counter enamel- The back of the plate is having counter enamel sifted over the fired base coat of black crackle. The supporting hand and fingers are all underneath the plate. The plate is held with its edge at a right angle to the sifter. The oxidized line of the sgraffitoed signature is not covered with the counter enamel, but it will be covered with soft flux before the piece is fired. *Photo by Bill Byers*

Doris Hall's Paisley Design- Top plate: 4" diameter, 18 ga copper, opalescent crackle over soft flux, transparent lumps in center, turquoise transparent border. Copper Bottom plate: 6" diameter, 18 ga paisley design. *Photo by Bill Byersy*

nates another, this mixed enamel (often called tweed) is put in a quart bottle and used for my counter. Two good siftings with the 60 mesh screen sifter are applied over the fired crackle, spraying with water before and after each sifting. The spraying between coats changes the color of the enamel enough to let you see where you have sifted. For the sifting on the back, I hold the plate with the center front of the piece on my fingers so none of my fingers are protruding from under the plate. I start the sifting around the edge, moving the plate in a circular fashion and then sift around towards my signature, but not over it. The center back has the signature covered with Thompson #426 leaded 80 mesh soft flux after two coats of counter are sifted on the back.■

TECHNIQUES FOR BEGINNERS

After you learn to apply enamels by sifting, wet packing, brushing on crackle and firing the base coats, you are ready for some of the simple techniques that require only a feel for color. You can design a piece by repeating one technique or learning to combine techniques. If you are enameling samples of the technique on 3" round or square 18 ga copper, one coat of crackle on the back is adequate. My beginners' course ended with Doris Hall's paisley design, which combined many techniques, and a simple cloisonné piece on a 1" disc of fine silver.

You need to plan the sequence in order to combine techniques. Have a design and color combination in your mind or on paper, at least the beginning of one. Either start with the design and decide which technique should be used for it or decide which technique you want to use and select the metal and the base coat of enamel for that technique. You need to select either a particular mesh enamel or a liquid form enamel, a transparent or opaque and a light or dark color. All these decisions are for the base coat of enamel on the front of the metal. As you develop the piece, you need to continue to make these same decisions.

For an abstract design with areas of color, using dry or wet stencils and maybe some sgraffito areas, your choice of the base coat influences the selection of the enamel layer or layers over it in subsequent firings, e.g., a transparent blue over an opaque yellow will give you a green. With enamel, as with oil painting, you usually can cover up an unwanted fired color with an opaque enamel in the following firing, especially if you have not fired a heavy coat. If you are planning on six or more layers of enamel, then each coat, except for the base coat, should be a thin application of enamel. To add other techniques to this piece designed with stencils, transparent and opaque sifted enamels and sgraffito, a fine line black drawing could delineate all or some of the shapes you have created. There are an infinite number of choices you could make. When I first learned to enamel, I would daydream about combining an assortment of the enameling techniques.

FIRESCALE

Bare copper, when fired, develops a layer of firescale. The longer and higher the firing, the thicker the firescale becomes on the bare metal. Sometimes, especially with a thick coat, it will flake off when the piece cools after the firing. The firescale layer expands and contracts with the firing as do metals and enamels. Most enamelists remove the firescale from the edge of a copper piece after each firing. When the piece has cooled, the firescale will usually chip off and could contaminate the enamel you are working with if you are sifting on another layer. With the firing, the color of the firescale changes after the second or third firing, from a rust tone to almost a black when a coat of flux is fired over it. I still use this firescale line for my signature on the back of my pieces. Judy Stone describes this line in her design technique of layering. In addition to drawing into the base coat to expose the copper, you can cover the line with sifted flux before firing. This will produce a flux line instead of a black oxidized line. You can also combine both the flux line and the oxidized line in a design by sifting the flux over some of the sgraffito and leaving some of it bare copper.

You can create a design with firescale. Dilute enameling gum with water 1:1. Brush the diluted gum in any shape on bare copper. Sift enamel over the entire piece, stand the piece on edge and tap it to remove the enamel from the un-gummed areas. You then have the enamel in the design you have brushed on the piece with the diluted gum. If the enamel did not adhere everywhere you wanted it to be, you can use a small sifter to sift that same enamel over those areas. Fire at 1500°F, only to maturity. If you over fire, the firescale will be thick and could flake off if you enamel over it. The enamel areas will be edged with firescale. This technique is most effective with a pale opaque or a light color transparent for the first firing. When the piece cools, file off the firescale from the edge of the piece and brush the face of the piece to remove any loose firescale. The second layer of enamel can be flux or any light transparent enamel sifted over the entire piece and fired. This firing seals the firescale and your piece is ready to be developed further with additional layers of enamel.

BUBBLING THROUGH

Fire a base coat of liquid form enamel on the back and soft 80 mesh opaque enamel on the front. Fire two siftings of counter on the back. Fire high a coat of medium fusing 80 mesh transparent enamel over all or part of the piece. The soft opaque enamel will bubble up through the transparent enamel.

COMBINING A DRY STENCIL WITH THE BUBBLE THROUGH

Instead of covering the whole piece with the transparent enamel over the opaque base coat, sift transparent enamel over the edge of a plastic shield held close to the piece. The closer to the piece you hold the shield, the straighter the edge of the sifted enamel. You can also use this technique to sift one enamel color over part of another in a plaid or stripe design.

FLOWING

Fire the base coats of any enamel you like. Use a small sifter to apply areas of three or four contrasting colors, opaque and/or transparent, close to each other. Some can overlap. Spray the piece with water until a little wetter than damp and tilt it in various directions to guide the flow of the enamels. If the enamels do not move, you can use water in an eyedropper to make the enamel flow as you tilt the piece. When the enamel design is to your liking, hold a piece of paper towel at the edge to draw off the excess water. Sift a light coat of soft flux over the piece to absorb the remaining excess water before you fire the piece to maturity.

WET STENCILS

Fire the base coats. Cut pieces of paper towel, wet them thoroughly and place them over a fired base coat. If the paper hangs over the edge of the piece, you can remove it easily after you sift on the enamel. If there is no overhang, use a pin and tweezers to lift off the wet stencil. If the copper piece has sloping sides, you might want to spray 1:3 diluted gum over the paper after it is positioned on the piece, then sift an enamel and spray the gum again. The sifted enamel should cover only the edge of the paper stencils. The piece is dry when the gummed enamel surface feels like sandpaper. Before you fire the piece, you can sgraffito the dried enamel with any sharp tool. If you are using only water under the sifted enamel, you can fire the piece when the enamel is just damp. The same washed stencils or new ones can be used for overlays in subsequent firings. The number of layers and firings depends on the design you envision. This technique is good for learning how one transparent looks over a number of transparent enamels.

PULLING THROUGH

First fire the base coats in an enamel of your choice. Using the Indian sand painting method, apply about four 80 mesh enamel colors in bands of color close to each other. First try enamel colors that are in sharp contrast to your base coat. Black and white are strong colors to combine. Put the point of a sharpened chopstick just beyond an outside band and drag the point through the other colors. If you raise the chopstick a little as you get through the last color, that color will end in a point. Pull through as many times as you like.

OVERALL LUMPS

Fire a base coat of flux on the front of a shallow plate and counter enamel the back. Spray the piece with diluted 1:3 gum and sift a light coat of white or a light color opaque 80 mesh over the whole piece. Place different sizes of soft fusing lumps over the whole piece and press each lump in place, which will move aside the sifted opaque enamel under the lump. Tiny lumps can be close together because they will not spread as much. The larger lumps will expand to cover more area. When the gum is dry, the piece is fired to maturity. If you place enough lumps on the piece they will almost touch each other when they expand and spread in the firing. The opaque enamel will frame each lump. After the lumps are fired, the piece should not be fired again upside down because the lumps might droop down to the floor of the kiln.

If your lumps are too large, put a few of them in a brown paper bag that is inside a plastic bag and put the bag on a scrap of wood. Bang on the lumps with a hammer. The lumps will scatter unless they are in a bag.

DORIS HALL'S PAISLEY DESIGN

This paisley design was a commercial enamel that Doris Hall and Kalman Kubinyi produced. When I began teaching, I used this design, with her permission, as part of my beginners' course because it combined so many enameling techniques. This enamel plate is shown on page **24**. The enamels were Thompson's old leaded 80 mesh and were fired around 1500°F.

PROCEDURE

Apply and let dry #767 Peacock crackle (now called Liquid Form Enamel) to the back of the plate.

Sift a good covering coat of #426 soft flux on the front over diluted gum.

Fire the piece flux side up, supported on a stilt. Front and back base coats are fired in the first firing. This method eliminates firescale forming in firing because there is no uncovered copper. All the subsequent coats are fired right side up also.

The second firing has another coat of crackle dried on the back and transparent #200 turquoise sifted on the front with the transparent and opaque lumps gummed in place. Push aside the turquoise enamel, about the size of the ½" red transparent lumps, from eight evenly spaced areas about ½" down from the edge of the plate. This exposes the flux base coat where the red lumps are to be placed with undiluted gum. The same is done with the transparent smaller lumps in the center of the plate. Then some small opaque lumps are just pushed down into the transparent enamel. With the tip of a palette knife, a little additional #200 enamel is placed at the bottom of the large red lumps around the edge of the plate. The added enamel forms a banking shelf to support each lump in the firing. With the gum dry and all the lumps in place, fire the piece until the lumps have smoothed down.

For the third firing, paint about a 1/8" wide line of separation enamel around each lump, leaving at least ¼" between the line and the edges of the lumps. The painted lines are connected with additional lines to form an overall design. The separation enamel, in a one-ounce bottle, with an oil base, has to be stirred thoroughly and a separate thin brush is kept to use just for it. The brush is cleaned with turpentine. When the separation enamel is dry, fire the piece. The painted lines, when fired, sink through to the flux and double the width.

For the fourth firing, apply a coat of the Peacock crackle over the entire front of the piece. When the crackle is dry, sgraffito with a pointed stick or tool, following the depressed lines made by the separation enamel, which exposed the flux base coat. Additional cross-hatched sgraffito lines should be done over the lumps. The crackle will break up and the lines will widen as you fire to maturity, or a little beyond, to complete the plate.

CLOISONNE PENDANT ON FINE SILVER

Fine silver cloisonné wire on a flat piece of 18 or 20 ga fine silver is the simplest material to use for cloisonné. You do not have to worry about burning up the fine silver wires as you do when you fire fine silver wires on copper. I fire in a heat-saturated kiln at 1500°F. The fine silver piece only needs to be cleaned in the kiln because it does not form a firescale coat, as does bare copper, when fired. My experience is only with leaded enamels, so the results may not be the same with unleaded enamels.

The first cloisonné piece my students created was a one-inch, 18 ga fine silver flat disc with a hole drilled for a jump ring.

MATERIALS

1" 18 ga fine silver flat disk with a drilled hole for jump ring
1' flat, fine silver cloisonné wire, .040" x .010". for the design
2 oz. soft flux, washed and dried. Thompson leaded 80 mesh soft flux #426 for the front base coat, or try an unleaded flux for silver
2 tsp. total of one or any number of 80 mesh transparent washed enamels to add color
1 brush full of thin crackle for the back base coat
1 tsp. uncut enameling gum for adhering wires to the silver piece before firing

TOOLS

Pair of pointed tweezers, with the points not very sharp, to bend the wires
Straight-blade bezel shears to cut the wires (see photo, page 10)
#1 sable liner brush to wet pack the enamels
1" 80 mesh sifter to apply base coat of flux to front of the disk
Carborundum stone for first stoning of top of the wires to remove any fired-on enamel
Scotch stone for final finishing on top of wires
A wooden board to put across a sink or set-tub for stoning the wires under water

PROCEDURE

Trace the disk about four times on 3" x 5" index cards.

Draw designs for the wires. For this first one, keep the wires at least 1/8" away from edge.

As you design, remember that the wires are to form cells in which the enamel will be wet packed with the brush. A wire must have a bend to be able to stand up on its thin edge. Select one design as your pattern.

Bend and cut the wires to your pattern. It is easiest to work with a 3" length of wire. When you cut the wires, the cut must be a true straight line for the wire to butt against another wire. Each wire must also be flat against the disk. Read Joseph Trippetti's method for bending wires to the pattern (see page **50**). Each piece, picked up with the tweezers, is placed on the cleaned disk with uncut gum.

When all the gummed wires of the design are in place and dry, sift the soft flux over the entire wire design area about one-third the height of the wire including an 1/8" on the outside wires of the design.

Fire the piece on a trivet at 1500°F until the enamel is holding the wires and the enamel looks almost white. If you use the medium fusing leaded flux, #1005, it will have a yellow cast.

For the second firing, wet pack soft flux with dilute 1:2 Klyr-Fyre around all the outside wires of the design and sloping down to the edge of the piece. Let dry and fire smooth.

When the piece is cool, paint a very thin coat of crackle on the back. When it is dry, sgraffito your name or logo.

Wet pack the transparent color into the cells with the tip of the liner brush. Push the enamel grains against the wires and then fill in the cell. Multiple thin fired layers are brighter than one thick layer of enamel. Add a drop of diluted gum to each cell. In the Swiss method, the enamel is not filled to the top of the wires. You can stop filling the wires and firing when the piece is to your liking.

At the sink, put a 3" square of chamois under the enameled disk and place both on the wooden board under a faucet in order to stone the top of the wires under running water.

Use the coarse Carborundum stone first, stoning in a circular motion. Then the Scotch stone is rubbed along the length of the tops of the wires. The side edge of the piece is either cleaned the same way or on a polishing wheel with a Bright Boy stone. At the wheel, you rotate the edge of the piece, keeping it moving against the spinning stone.

A sterling jump ring is set in the drilled hole, and the piece is complete.

You can also design the cloisonné piece with bent wires without drawing a design. Just bend pieces of wire and place two bent pieces on the silver in the center of the disk. The end of one piece butts against a length of the other piece. Then you add pieces to make enclosed cells, designing with the bent and cut pieces. If you are a jeweler and intend to bezel the enamel, it will be easier if your wires are 1/8" in from the edge. A pair of tweezers and your fingers are all you need to bend the wires along with a feeling for design.■

PROFESSIONAL ENAMELISTS' TECHNIQUES
CLOISONNÉ AND PLIQUE-À-JOUR

CLOISONNÉ BEADS OF FINE SILVER

LINDA CRAWFORD, *Linda Crawford Designs*

Linda Crawford was born in Corona, CA. She has studied oil painting, drawing, jewelry fabrication, ceramics, weaving, art history and cloisonné enameling. In 1995, she turned to cloissoné enameling full time. She teaches workshops on cloisonné enameling at the California Institute of Jewelry Training in Sacramento. Her cloissoné jewelry is shown and sold throughout the United States. With two other jewelers she maintains a working studio gallery, Mendocino Jewelry Gallery, Mendocino, CA. Her themes blend the energies of spirit and nature.

I have been making my cloisonné beads since 1996 in a variety of methods with fine silver and wires of fine silver or 24k gold for the designs. My method of fusing two domed halves of fine silver requires fewer metalsmithing skills than the other methods I use, which is why I have chosen to explain this one. Although I currently create my beads the way I describe here, the door is always open for experimentation.

I make 1/2", 3/4", and 1" beads. The beads usually have a cloisonné design on them and require anywhere from six to twenty firings. A rule of thumb is the smaller the piece, the thinner the gauge of metal. For a one-inch bead I use 20 ga fine silver. For the cloisonné wires, I purchase 20 ga fine silver, round wire and 20 ga to 26 ga in fine gold. The wire is rolled to the size I want to use. I start by drawing the design for the cloisonné wires but I do not cut and form them until I have made the bead. I finish the bead by inserting a short piece of sterling tubing into the hole I have drilled and flanging the ends of the tubing. This protects the edges of the enamel. If I use gold cloisonné wires, then I make gold tubing. I mostly use my 9"w x 9"h x 12"d electric kiln with a pyrometer.

To support the bead or beads while applying the cloisonné wires and for the firing, I make a bead jig using an 8" square of stainless mesh. Two opposite sides are bent up 2" high, which leaves a 4" x 8" bottom. I use a wire cutter to remove every other wire from the top edge of the two bent up sides, which gives slots that will support a wire that goes through the bead. I can fire three beads at one time with this jig. For wire going through the hole of the bead, I use a 4-1/2" length of clothes hanger wire, which I have coated with Amacote, a firescale inhibitor. My firing setup is a 6" stainless mesh (the corners have been bent down 1") with a square piece of mica on top of it to catch any dripping. The 4" x 8" support that holds the beads on the dry coated wire is placed on top of the mica.

To make the bead, I cut out two circles, drawn with a circle template, with a jeweler's saw. Alternatively, a disk cutter can be used or the circles can be ordered precut. The silver disks are annealed on a charcoal block with a propane/oxygen torch, quenched in water, dried and

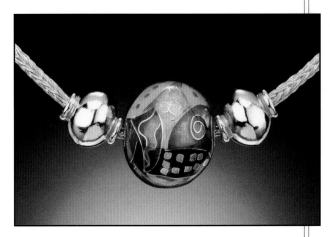

Center bead, ¾" diameter, cloisonné on fine silver. Chain is hand woven fine silver. All metal hand fabricated. *Photo by Hap Sakwa*

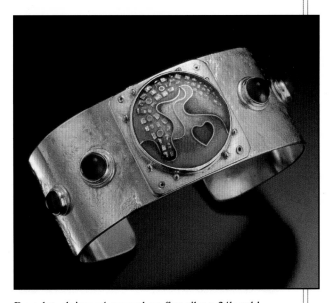

Bracelet, cloisonné enamel on fine silver, 24k gold wires and foil, setting: 22k gold bezels, sterling.

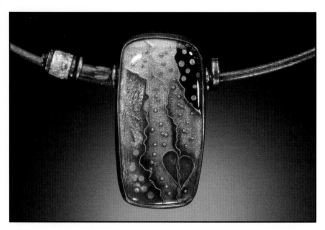

Neckpiece, cloisonné enamel oblong bead on fine silver, 24k gold wires, gold foil, ground granules, setting: sterling, 22k gold. *Photo by Hap Sakwa*

"Amulet" pendant, cloisonné enamel on fine silver, 24k gold wires, snowflake agate, citrine, *mokume gane*; setting: sterling, chain: hand woven fine silver. Bead spins on tube rivet.

then domed in a dapping block with punches to begin forming the bead. I gradually move deeper in the dapping block and anneal between each depression. This process continues until I have two halves that make a sphere. On the underside of each dome, I center punch and drill a hole that will just accept the clothes hanger wire and the tubing that will line the hole. I clean up the burrs in the drilled holes with a round needle file. I make a small handle with masking tape at the top of each piece to make it easier to hold the pieces for sanding. The edges of the two halves are sanded on a steel plate with 280 grit sandpaper followed by a 400 grit sanding stick. The goal is a perfect fit. The pieces are cleaned with a water/ammonia mixture using a soft toothbrush, rinsed well and dried before fusing them together with the torch.

First I use a round burr to make a small divot in the charcoal block. The divot will support the bead. I place the bead pieces on the charcoal block and spray them with Cupronal flux, which is specifically for silver. Using a soft flame, the flux on each piece is dried to a white powder with a soft flame on my Meco Midget torch with a #3 tip. Then the two pieces are placed one on top of the other to form the bead. You must be careful not to melt the bead. Heat the bead evenly, both the top and bottom. The silver will start to glow red. As the bead becomes almost shiny, concentrate the heat on the joint between the pieces. You will see the silver flow into the joint. Rotate the block so that you can fuse (flow the metal) all the way around. When the fused bead is cool, I usually texture it with a bud burr or diamond impregnated bits. Check again that the seam is completely fused. If all is good, the bead is pickled in Sparex 2 solution just long enough to clean off the soldering flux.

Next, rinse the bead and put it in a light solution of ammonia and water so that the solution goes inside the bead. Rinse the bead thoroughly under running water; making sure the inside of the bead is clean. Then burnish the bead with a glass brush to bring up the shine. While the bead is drying, cut a 4-1/2" piece of straight coat-hanger wire and dip the wire in Amacote or another firescale inhibitor. When both are dry, slide the bead onto the Amacote coated wire.

The adhesive I use is CMC (sodium carboxymethylcellulose), a cellulose gum that completely burns off during the firing. CMC comes in dry form, which allows me to adjust the viscosity. You can experiment to find your favorite mixture. CMC has a limited shelf life after it is mixed with water so I only mix the amount I need for the next few weeks. I mix it by stirring the water while sprinkling in the CMC. A small enamel sifter works well for sifting the CMC into the water. CMC comes with directions, but I eyeball it for the solution. The enamels are washed with tap water and given a final rinse with distilled water. I let the flux dry so that I can sift it on the gummed bead as a base coat. The gum is a thin coat of diluted CMC that is sprayed on. I previously used Schauer's

#2A flux but since that is no longer available I use the Japanese, leaded Ninomiya N1 flux for the base coat. I apply wet enamel inside the larger beads with either a 2/0 or 4/0 rounded tip, fine sable brush, whichever size fits in the hole of the bead. This base coat is fired at 1475°F for 1-1/2 minutes. After the piece is cool, I apply another thin coat of flux, firing for two to three minutes to maturity.

The bead now receives the cloisonné wire. I purchase dead soft, 20 ga, 24k round wire and fine silver round wire that I draw down and roll to the rectangular size I want. One of the sizes I use is 29-30 ga x 17-18 ga rectangular wire made with the rolling mill set at .006. The wire is annealed during and after the drawing and rolling. The annealing can be done by placing a carefully coiled roll in the kiln.

The wires are bent to the design. Each wire also must be fitted to the curve of the bead; it is held in place with a thick CMC mixture. I use tweezers and my fingers to shape each piece. The cut and shaped piece of cloisonné wire is dipped in the CMC and placed on the bead that is on the coated wire. Only the bottoms of the cloisonné wires that touch the bead need the CMC: do not flood the piece.

There are two ways to put the wires on the bead. One way is to cover the whole bead with wire, dry it, and then fire, or put the wire on in sections. I start placing the wires at the top section of the bead. If one of the wires needs to be repositioned before it is fired, I dip the tip of my brush in distilled water and use it to move the wire. Each wire should be standing straight up on its thin edge. Let dry. Fire the bead for two to three minutes at 1475°F. An indication that the wires are adhered to the bead properly is when a thin glossy line appears at the base of each wire. Let the bead cool and adjust any wires that have not adhered properly. Be careful not to press the wires too hard because they could collapse.

The selected enamel colors are washed with distilled water and kept wet in small covered plastic or glass jars. Wet packing with just the tip of a fine, sable brush, I pick up and place the enamel in the cloisons. After one side of the bead is completed with the enamel packed to the top of the wires, the excess water is drawn off with the edge of a tissue. A tiny drop of watery CMC is applied to the damp enamel. Continue packing the enamels until the entire bead is covered with one layer of enamel. The bead is let dry and then fired for 1-1/2 minutes at 1475°F. I under-fire the first layer. You will have to experiment with your kiln for temperature and firing time. The second firing will be for two minutes and subsequent ones built up to three minutes, which helps to ensure that the molten glass will not pool on the bead and cause an uneven surface. The layering and firing continues until the cells are filled to the top of the wires.

The enameling finished, the bead is stoned under running water. I first grind the entire surface with a 120 Alundum stone until all the

Necklace, cloisonné enamel bead on fine silver, 24k gold wires and foil, amethysts, aquamarine gem stone, setting: sterling, 14k yellow gold.

Earrings, cloisonné enamel on fine silver, 24k gold wires and foil, setting: sterling, 22k yellow gold bezels.

wires show and the surface starts to be smooth. Then I rinse and clean the bead with a glass brush until all the small dust particles are gone. If there are any glossy depressions, those areas are re-enameled and fired for about three minutes at 1475°F. This time I stone with a 200 Alumdum stone. If there are still glossy depressions, then the filling and firing and stoning continues until they all disappear. The piece is finished by polishing it.

The bead is placed on a dop stick with wax to hold the bead easily for polishing. An enamel is always stoned or polished under or with water. I start with a 400-diamond lap wheel with a water attachment. After polishing the exposed section of the bead, I put it in the freezer for five minutes in order to remove the dop wax. The wax should just pop off the bead when you take it out of the freezer. Let the bead warm a little in your hand before placing it under running water. If there is a piece of wax that is stubborn, put the bead in the freezer again for a few minutes. When all the wax has been removed, dry the bead with a soft cloth and repeat the polishing until you have done the entire surface.

If I want to hand finish the bead, I start with 325 wet/dry sandpaper, then go to 600 and finally to 1200 grit. The bead is cleaned, under water, with a glass brush and dried before I flash fire at 1500°F for one to two minutes. The firing time depends on the size of the bead. This final firing will put a high gloss finish on the bead and will seal off any "pores" that are on the surface of the enamel. I use this finish specifically for jewelry that will be worn because I feel it protects the enamel from dirt and other contaminants.

The last step is the insertion of a piece of sterling silver tubing. I usually make my own tubing, but you can purchase tubing. Specify that it be the size you want and thin walled. The hole in the bead is made slightly larger than the tubing for the for the tubing to slide easily in and out of the bead. Make certain that the bead is centered on the piece of tubing. Put a scribe into an end of the tube and swing it around to flare out the end of the tube. Do the same to the other end of the tube. To complete the rivet, I use the rounded end of a small ball peen hammer. Place a tubing end on a metal plate and gently tap first in the center of the tubing and then around until the edge of the tube is curled over the bead. Complete the rivet by doing the other end of the tube.

The tube may be burnished with a burnishing tool to bring out a shine. I use a red rouge impregnated buff on the wheel to put a shine on the wires and the tube ends. After buffing with the red rouge, I soak the bead in warm water with a small amount of ammonia and dishwashing liquid and then clean it with a soft toothbrush. I draw a cotton string or a pipe cleaner soaked in the cleaning liquid through the center of the tube. When all traces of rouge are gone, I soak the bead in warm, clear water, shake off the water and let the bead dry.

The word bead comes from the Anglo-Saxon root word "bebe" that means "prayer." "Bidden" means "to pray." My fondness for beads began with the rosary prayer beads when I was a young child. My beads are not always meant to be worn, but sometimes to be used for meditation ora tool in prayer.■

Cloisonne With 24K Gold Wire and Gold Foil Basse-Taille

ERICA DRUIN

A second generation enamelist, Erica Druin was named a 2004 Niche Awards Winner and First Place Winner, Lapidary Journal's Bead Arts Award Competition 2003 and 2004. Juried shows and exhibitions: Philadelphia Art Museum Craft Show, Washington Craft Show, ACC Baltimore Winter Craft Show, Palm Beach Fine Craft Show, SOFA New York and Chicago, Enamel Society's 9th International Juried Exhibition.

Erica Druin (right) with her mother, Marilyn Druin.

"Forever" brooch/pendant, 2.5"x 2.5", 24k gold cloisonné/basse-taille/guilloche, enamel on fine silver, setting: 18k and 24k gold. *Photo by Ralph Gabriner*

As a second-generation enamelist, I feel blessed to be continuing the tradition started by my mother over 40 years ago. Some of my earliest memories are of spending time at craft shows and museums. Little did I know that words like "cloisonné" and "basse-taille" weren't part of every seven-year-old's vocabulary!

Although enameling was always a strong presence in my life, it wasn't until my mother was diagnosed with cancer that I actually made my first piece. I took a vacation from my job in hospital administration to work side-by-side with her in her studio. Near the end of my time off, I called work to let them know I'd be gone a few extra days. It was then that my mother and I both realized that I was truly enjoying enameling and not just doing it to humor her. After I completed my first small pendant, I knew I wanted to do more. My mother and I had great plans for spending more time together in the studio. Unfortunately, her cancer did not allow that.

Prior to her passing away, my mother had been planning a collaborative exhibition with metalsmith Michael Good. After her death, Michael and I decided that the exhibition should still take place. Michael asked me about a piece that was to be a part of that show: a scalloped bowl he had formed for my mother to enamel. I found the bowl on the corner of her workbench. It had no color beyond its base coat of red enamel on the outside and white on the inside. As I sat staring at it, I thought about how much I missed my mother. I became determined to finish this piece she'd only begun. I had no idea where to start. Only a few long cloisonné wires were in place, but upon closer inspection I noticed one small swirled wire. This small swirled wire then seemed to direct me to the ultimate design of the piece. When I fired the piece to fuse on the wires, gravity and the 1500°F kiln had their influence on the three-dimensional form. Only one wire fell off during that first firing, the original swirl placed by my mother. It was as if that small swirl had served as my mother's presence guiding me in the design of the piece. I saw its departure as a sign that the world of enameling was now mine to continue in, and that's exactly what I've done.

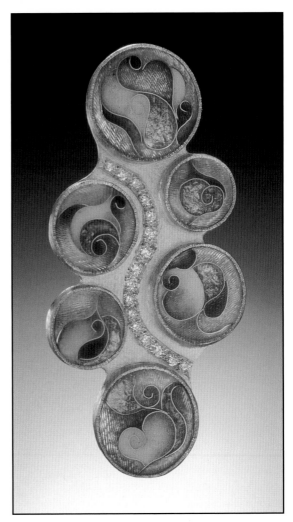

"Reflection" brooch/pendant, 2.625" x 1.25", 24k gold cloisonné/basse-taille, guilloche, enamel on fine silver, setting: 18k and 24k gold, pave diamonds. *Photo by Ralph Gabriner*

"To Everything There is a Season" memorial plaque, 2.5" x 2.5", 24k gold, cloisonné/basse-taille/guilloche, enamel on fine silver and 24k gold. *Photo by Bob Barrett*

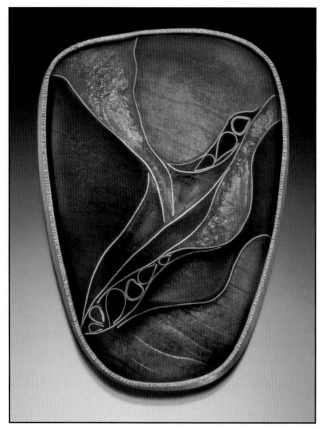

"Colors" pendant, 1.625" x 1.125", 24k gold cloisonné/basse-taille, enamel on fine silver, setting: 18k and 24k gold. *Photo by Ralph Gabriner*

I use fine silver as the base metal for my enamels. Any doming is done with an hydraulic press before sawing and filing to the desired shape. The piece is cleaned by heating it in the kiln and burnishing it with a glass brush under running water. The base coats are then applied. I begin by using a deer tail brush to paint Klyr-Fyre on the back of the piece and sifting a medium coat of counter enamel; I am careful that none gets on the front of the piece. My counter enamel is a combination of leftovers of all my enamels. The piece is dried, fired and allowed to cool.

The front of the piece is then painted with Klyr-Fyre. A thin coat of N-1 flux for silver is sifted on, and the piece is dried, fired and allowed to cool. A second medium to heavy coat of counter enamel is then applied. I maintain my kiln at a temperature of 1500° F. I turn the kiln on high for firing and then turn it back down after the firing. I never put down the fork when a piece is in the kiln so that I cannot become distracted, forget that a piece is in the kiln and potentially over fire it.

Once the base coats are completed, I am ready to apply my cloisonné wires. I never sketch or draw my design. Instead, I create the design as I bend the wires, which are annealed before I use them. I use 24k gold cloisonné wire. The wires are .040" in height, and their width can be anywhere from .005" to .015", depending on the thickness of the line I desire. Klyr-Fyre holds the wires in place. When working on a three-dimensional form, I sprinkle lily root into the Klyr-Fyre for added strength. Once the wires are in place, and the Klyr-Fyre dry, the piece is fired. I will often sprinkle a small amount of flux onto the piece before firing to make it easier for me to see when the firing is complete.

After the cloisonné is completed, I add textured metal to create areas of basse-taille. I use 24k gold foil or gold sheet that I roll very thin. I typically texture the gold sheet by laying it over a template of textured metal and putting it through the rolling mill. In order to cut the textured metal to the appropriate shape and size for the piece I will either photocopy the piece or place paper over the piece and gently make a rubbing of the cloisonné design. I can then cut out the area I want the gold to fill and use the paper as a pattern for cutting the metal. Once the metal is held in place with Klyr-Fyre, I put a small amount of flux over it prior to firing.

I use old Thompson enamels as well as Japanese transparent leaded enamels in my work. Enamels are 80 mesh and sifted with a 200 mesh sifter. I place a coin in the sifter with the enamel to create added friction and put an extra pan on top of the sifter to decrease the enamel dust in the air. Sifted enamels are stored in jars. Plastic shot glasses are used for the enamels when I am working. The shot glass is filled half-way with cold tap water to wash the enamel. Approximately one-third of a plastic teaspoon of enamel is added to the glass. A wooden chopstick is used to stir the enamel. After the enamel settles, the water is poured off the top. This process is repeated until the water pours clear.

I wet pack my enamels using various size natural fiber brushes. My initial layers of enamel are typically light colors so that as the color scheme of the piece unfolds I have the greatest amount of flexibility. The light colors can always be made darker if the design calls for it, but a dark transparent color cannot be made lighter. Paper towel is used to soak up the excess water once the enamel has been placed, and the piece is allowed to dry completely before firing. I do a great deal of shading in my pieces, which involves many layers of enamel. Sometimes I feel like I am adding a mere three grains of enamel at a time in an effort to achieve the shading I want. A typical piece of mine may be fired anywhere between ten and fifty times.

When filling the cells with enamel, I make sure to protect the cloisonné by bringing the enamel at the cell edges to the level of the top of the wires. I do not necessarily bring all the enamel to that same level; I often will leave the enamel shallower within the cell. I like the added three dimensionality this variation brings to certain pieces.

Once the enameling is done, I use a wet grinding machine to clean the edges of the piece as well as the cloisonné wires. When I am sure the cloisons are completely uncovered, I clean the piece with a glass brush under running water. Armour Etch is then used to bring the piece to a uniform matte finish. It is once again cleaned under running water with a glass brush and then various wet/dry papers (320, 400, 600) are used to bring the piece to the desired finish.

As I work, the design of the piece is constantly evolving. Under the influence of gravity and the heat of the kiln, the cloisons may ultimately move to a different place than I originally planned. The shadings of color created in the enamel may lead in an unexpected direction. I don't fight these changes; I embrace them as the ultimate wonder and excitement of enameling. I truly believe the piece itself will guide its development and tell you what it's meant to be if you are open to it. And if it isn't perfect, that's okay: I don't really want to create the perfect piece right now because if I did, I'd be done enameling.■

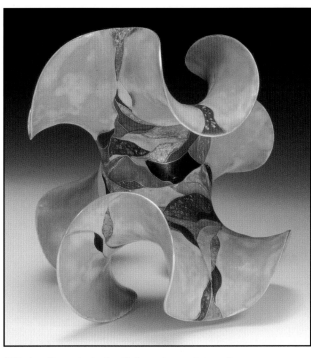

"Waving Boundaries" collaborative sculpture by Michael Good and Erica Druin, 4.5" x 4.25" x 2", 24k gold cloisonné/basse-taille, enamel on fine silver and 24k gold. *Photo by Ralph Gabriner*

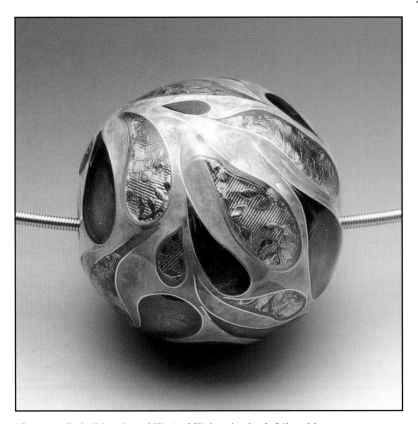

"Summer Rain," bead, 1.125"x 1.125", hand-raised, 24k gold cloisonné/basse-taille enamel on fine silver and 24k gold. 2004 NICHE award winner. *Photo by Richard Goodbody*

Gold Cloisonné Wire On Fine Silver

FALCHER FUSAGER, *Magick/Fusager-Demski Design*

Falcher Fusager, born in Denmark, moved to the United States in 1973. While working as a street artist, he gradually transformed his art to enamel cloisonné jewelry. He is self-taught in all areas of his work.

Susan Demski, born in Chicago, IL, moved to California in 1972. She received a BA in Art, emphasis in enameling, from San Diego State University in 1976. She was a commercial interior designer until 1985 when she focused on enameling. They formed their company, Magick, in 1988.

Susan Demski and I have developed a cloisonné technique for our jewelry, which ranges in size from 1/4" diameter to about 4" x 5". Our pieces are enameled on fine silver with gold cloisonné wire and then set in karat gold. For the enameling, we use 22 ga for earrings and 20 ga for most other pieces. We prefer the fine silver because it is a bright, neutral color that does not oxidize. Our enamels require from 15 to 20 firings.

We start with an exact color drawing. We either cut the shape from the drawing with table shears or order the pieces precut to our specification. The fine silver piece is cleaned with sandpaper and then slightly domed.

Next the front is usually engraved, as in basse-taille, in continuous lines for reflection or in intricate designs as a decorative element. We use a Dremel electric engraver. Sometimes the design is stamped or rolled in the metal.

The enamels we use are 80 mesh, leaded, in both transparents and opaques. The enamels are washed with a final rinse of distilled water. The counter enamel is purchased already mixed. It is sifted dry on the back, about 1/2 to 1 mm thick. Firings are at 1450°F. The base coat on the front is usually medium or hard flux that is sifted on dry and fired to maturity. If this sifted fired layer does not completely cover the metal, another application of flux is fired. Then come the gold cloisonné wires.

We purchase fine gold wire in various gauges, from 28 ga to 14ga, which are rolled to different thicknesses starting at .002". We use

"Creator VIII" pin and pendant, 5" w, cloisonné, 24k gold wires on enameled fine silver, setting: 18k gold, tourmaline, diamonds.

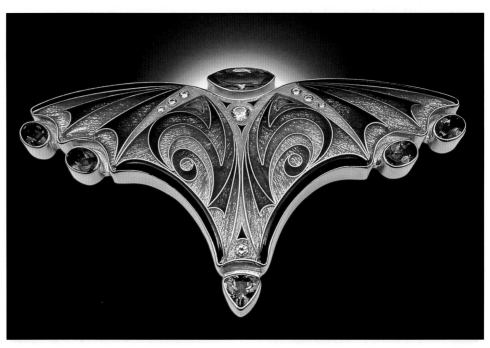

various size wires in each piece. The thin wire is used for small details. We bend and cut the wires to the design with fine tweezers and tiny scissors. The wires are placed on the fired flux coat with uncut Klyr-Fyre. When the gum is dry, the piece is fired until the wires are embedded in the flux.

We have three kilns, all about 12" x 12" x 7" with a pyrometer and a rheostat set at 1450°F. We fire from one to twenty pieces at a time. Each piece is placed on a high-fire steel trivet, and the trivets are placed on a steel rack. The rack is supported by two spaced bricks on the floor of the kiln.

The colored enamels are wet packed with distilled water and applied in a thin layer with a small scoop and a needle under magnification. Special care must be taken to pack the grains tightly because enamel liquefies on the surface first, trapping any air left between the grains during build-up. Besides creating problems during the final polishing, trapped air bubbles lower the brilliance of the enamel. The key to success is thin layers and tight packing.

Pieces can be tightly packed by lightly tapping the side of the piece with a tool, but we use the vibrating edge of an electric engraver held to the side of the disk. After vibrating the piece, we soak up any excess water with a piece of paper towel. The enamel piece is dried on the top of the kiln and then fired. This process is repeated for each layer. The number of layers possible depends on how thinly each layer is applied and the height of the cloisonné wire.

The layering of the enamels enables shading, which can greatly accentuate the color and give a great feeling of depth. We shade by packing grains of various shades of the same color next to each other. In this fashion, you can start with clear enamel on one end and finish with very dark enamel on the other end to create a high dynamic range. In subsequent layers, the different hues are shifted slightly to overlap the colors below. This shading usually requires from five to ten layers. When the enamel reaches the top of the wires, the piece is ready for finishing.

The final finishing of the enamel is done first with an aluminum oxide #80-100 belt on an expandable rubber drum and polished with a #600 belt. The drum is 8" diameter and 3" wide. It is the regular stone-cutting type setup with water. At this point, there is a choice. If there are no air bubbles in the enamel, the enamel can be polished with cerium oxide for a wonderfully smooth surface or it can be flash fired at 1500°F just until the surface is glossy. We prefer a flash firing because it is a more durable surface and enhances the brilliance of the enamel.

And that's it. Your gem, the precious enamel, is now ready for its gold setting. ■

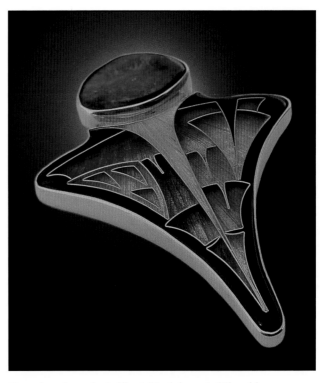

Brooch and pendant, 2" x 1.5", cloisonné, 24k gold wires on enameled fine silver, setting: 18k gold with opal.

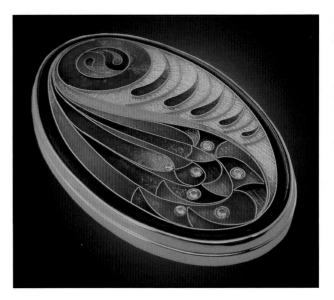

Brooch, 2.5" x 1.5", cloisonné, 24k gold wires on enameled fine silver, setting: 18k gold with 18ct diamonds.

"Energy," 3.5" h, cloisonné, 24k gold wires on enameled fine silver, setting: 18k gold, drusy, diamonds.

"Egyptian Collection" pendant, 2" h, cloisonné, 24k gold wires on enameled fine silver, setting: 18k gold, garnet, diamonds.

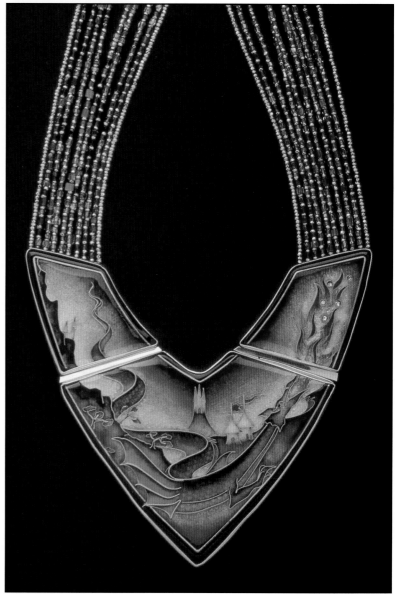

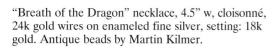

"Breath of the Dragon" necklace, 4.5" w, cloisonné, 24k gold wires on enameled fine silver, setting: 18k gold. Antique beads by Martin Kilmer.

Cloisonne Shaded Flowers

SANDRA KRAVITZ

Sandra Kravitz, educated as a botanist at London University, England, apprenticed to a Yemenite silversmith in Israel and then studied enameling in the United States. She is the leading designer for the largest Judaica manufacturer in the United States. She teaches enameling at the Craft Student League, Manhattan, and in Great Neck, NY. She has been President of the Enamel Guild North/East, since 1999.

An enthusiastic gardener, I was trained as a botanist. I love using flowers as the subject or inspiration for my enamels, which are usually on a small scale. My work is jewelry and mezuzahs ranging from 1.25" to 3". I take a photograph of the flowers at the same time that I draw them so that I'll have the colors available when I enamel. Although I work from nature, the design later may become abstract.

I work mostly on fine silver that is domed in a hydraulic press or a wooden dapping block. With the hydraulic press I use 24 ga or 26 ga fine silver; I use 22 ga with the dapping block. The press domes uniformly causing less stress and warpage. The fine silver is cleaned with detergent and a toothbrush to remove grease. It is next burnished in a tumbler with steel shot and soap medium for a high shine and then rinsed to remove all the soap.

I have three sizes of kilns. My favorite is an old Norman kiln that is only 5"x 5"x 5" with 3.5" of insulation. It comes up to temperature very quickly and retains the heat. All my kilns have electric controllers; I usually fire at 1450°F.

I use leaded enamels: Thompson, Japanese Ninomiya or Soyer at 80 mesh. Under red or orange transparents, I usually fire a layer of gold foil for more glowing colors and then a thin coat of Ninomiya N1 flux. To ensure clarity in the transparent enamels, I remove the fines, which cause opacity, either by sifting them through a 350 mesh sieve or by washing the enamels in distilled water about six to eight times. For wet packing the enamels, I remove most of the water after washing the enamels and add one drop of Klyr-Fyre. When the enamels have dried after application, the Klyr-Fyre prevents the enamel from moving, which is important for fine shading. Other than counter enamel and a base coat, which I sift, I apply enamel by wet packing with sable water color brushes from number 1 through 00000.

To counter enamel, I gently brush a little undiluted Klyr-Fyre onto the silver with a ½" flat sable brush, making sure not to produce any bubbles, and then sift Thompson's T121 emerald green; fire; resift and re-fire for a second coat. Before enameling the front, I lightly burnish the silver with a glass brush under running water. A base coat of Ninomiya Flux N1 is sifted on the front by the same method as on the front and fired. If I am doing cloisonné, this flux is fired just to orange peel.

"Arrangement in Grey & Black II" brooch, 2.25" x 2", cloisonné enamel on fine silver, fine silver wires, setting: fabricated, sterling silver. *Photo by Alan Kravitz*

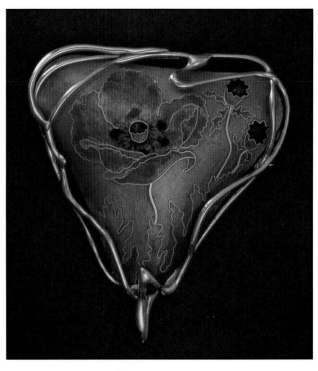

"Poppy" brooch and pendant, 2.375" x 2.625", cloisonné enamel on fine silver, fine silver wires, setting: 18k cast gold (wax model by artist). *Photo by Alan Kravitz*

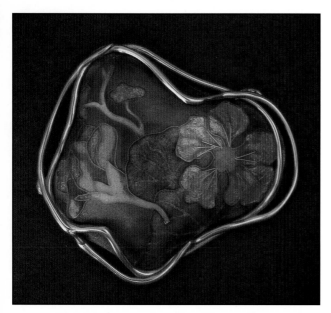

"Nasturtiums" brooch, 2.5"x 2.25", cloisonné enamel on fine silver, fine silver wires, setting: 18k cast gold (wax model by artist). *Photo by Alan Kravitz*

"Daffodil 2" brooch, 1.5" x 2.625", cloisonné enamel on fine silver, fine silver wires, setting: cast sterling silver (wax model by artist). *Photo by Alan Kravitz*

When the cloisonné wires are fired in place, I fire to maturity. This way it is easy to see whether the enamel is melted down enough to hold the wires in place. My cloisonné wires are all 0.010" x 0.040" or 0.060" deep. The depth depends on how many layers I will need. If using foils, the deeper wire is better. If I want a tapered line of wire, I cut a strip of 26 ga fine silver the depth of the other wires and step roll it. Having shaped the wires to my design, I dip them in a little Klyr-Fyre and position them on the fired flux layer. If there are many small wires like the stamens or veins, I place and fire the larger wires and then the smaller wires.

I use three types of shading, two with only transparents and one with an opaque base coat. An example for an opaque is a white flower. A white flower needs a base coat of medium fusing white fired and then a coat of opalescent white. If you look at a white flower, it contains many colors, for example, green, yellow, gray. The shading is applied with transparents or opalescents. These first layers will be brighter than desired but they will be muted by an opalescent top coat. Care must be taken not to build up too thick a layer of these colors, especially close to the cloisonné wire. If the enamels creep up the wires in the firing, the wires will be uncovered during the finishing process and create a hard, bright line. Fire after shading, and if needed, another layer of shading can be applied and fired. Finish with thin layers of opalescent white, firing in between each layer until all cloisons are filled.

To shade using only transparents, a range of shades of the same color are chosen. These colors are mixed with equal parts of soft flux to ensure they do not get too dark. The dry powdered enamel is mixed with up to 50 percent dry soft flux, and then the mixture is washed.

An example of Thompson leaded transparent colors for red poppy petals, from lightest to darkest, is:

Hazel #T112
Burnt Orange #T531
Garnet #T1066
Raspberry #T676

Always be certain that there is sufficient area of the lightest color because you can never make a transparent area lighter; only darker. Build up with these colors until you are satisfied and then use the garnet over all for the last layer to be fired.

Another shading method with transparents that I use for leaves is to first fire a layer of the palest transparent green, for example, Ninomiya N45 or Thompson Palm and then shade, according to the leaf color, with darker greens or pale gray-blues. After firing the shading, I fill in the cloisons with the pale green, mixed with soft flux if necessary. Never fill with only flux, since a thick coat of flux will cause cloudiness. I suggest that you make a few test pieces of the possible shadings you are thinking of using.

Flower stamens, which are usually very thin, can be made in several ways. If the stamens are light, they can be formed with fine silver wire, either cloisonné wire or 30 ga round wire. I make darker stamens from an enamel ground with an agate pestle and mortar or from an enamel with the fines removed by sifting the enamel with a 350 mesh sieve. To apply and manipulate these very fine lines, I use a combination of a 00000 sable brush and a needle. I make a thick ink by using the back of a plastic spoon to mix underglaze black, Thompson P3 powder, with drops of water; I use it with a pen for black stamens. Other thin lines such as veins or tendrils can be enameled in the same way as the stamens. Thompson's P1 and P3 are the same except that the P3 is denser. I use the P3 as both underglaze and overglaze.

When the enameling is finished, the surface needs to be ground. I do the grinding on a wet expanding drum sander with a 600 grit wet and dry belt until the enamel is smooth and all the wires are polished. After cleaning the piece thoroughly with a glass brush under running water, I fire it again to finish.

I have explained how I enamel with flowers as my subject, but I hope that those of you who enamel will find it helpful for other subjects.■

"White Dogwood 1" brooch, 2.5" x 1.5", cloisonné enamel on fine silver, fine silver wires, setting: 18k cast gold (wax model by artist). *Photo by Alan Kravitz*

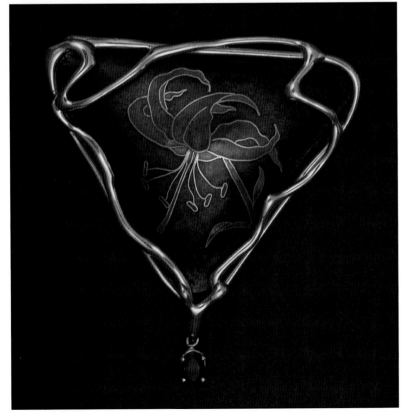

"Tiger Lily" brooch and pendant, 2.25" x 2.375", cloisonné enamel on fine silver, fine silver wires, setting: 18k cast gold (wax model by artist). *Photo by Alan Kravitz*

Cloisonné on Fine Silver with 24K Gold Wire

MARGARET LANGDELL

Margaret Langdell found her calling when she discovered some enamels while working at a craft center. Enameling mixes the beauty of art with the rigor of science. Working for Josh Simpson improved her knowledge of glass and its possibilities. She left to attend the North Bennet St. School's jewelry program. She teaches enameling at the Brookline Arts Center in Massachusetts and consults with museums and collectors about their enamels.

Margaret Langdell

I enjoy making complicated necklaces with three to ten panels joined together to tell a story, often a myth. The panels are rarely over 2.5". Usually each panel is set in a stepped bezel, and the bezels are linked together. I primarily enamel on 26 ga fine silver because it gives me a "white" surface on which to paint. The 24k pure gold cloisonné wire does not react with any of my enamels. When I use the 22k NT gold wire, a gold and silver alloy, some of my opalescents turn yellow.

I die cut, then dome, the fine silver pieces with the assistance of Anne Larsen Hollerbach and her hydraulic press. The fine silver does not oxidize, and the 26 ga is less likely to stress the enamels. With such a thin layer of silver, the high ratio of glass to metal is less likely to crack. The doming helps to reduce sagging while the remaining flange reduces warping during the twenty to thirty firings required for each panel.

I cut out my domed shapes with a jeweler's saw and then texture the top surface with a large bud burr in a flex-shaft. The back is enameled first with three thin applications and firings of counter enamel, often Ninyoma LT60, a teal blue. I like the blue because the counter enamel shows in the stepped bezels I most often use. In addition, it does not react with fine silver and it beautifully shows gold lettering. Once the counter has been completed, a thin coating of Ninyomia's N3 flux is sifted over a thin coat of Klyr-Fyre. The flux is no more than two grains deep. When the piece is dry, it is fired to orange peel stage. Ninyomia's flux comes pre-sifted without fines, and I do not wash it. If applied in a thin coat, it fires with perfect clarity.

For my larger pieces, I make a test strip of the enamels. I fire horizontal lines of opaques and opalescents. The second firing is stripes of transparents. I want to ensure good color matches, avoid any chemical reactions and make sure that fusing temperatures are fairly uniform. Although Japanese colors rarely have these problems, I also like using the old Thompson leaded enamels and don't want any surprises.

I make my own 24k cloisonné wire by running 24 ga round wire through the rolling mill to approximately 36 ga by 1.5 mm high. The wires are bent into shape with tweezers on a flat drawing and then placed on the panel. The drawing is kept inside a heavy three-ring binder transparent sleeve that protects the paper from the tweezers. I bend each wire on the drawing in the sleeve and then place it on a photocopy that has double-stick tape on it in order not to cut duplicate wires.

After all the wires are bent, I gently pinch the top of the wires with round nose pliers to fit the curvature of the panels. This procedure is a modification of forging; it elongates and curves the wire pieces. I also clip the bottom corners of my cloisonné wires to prevent unwanted spaces across the surface of the design. I apply the wires with lily root powder. I put a couple of ounces of tap water into a small dish and next sprinkle a dusting of the powder across the surface of the water. Within a few minutes, it becomes a gelatin and is ready to use. Usually, I start by applying just a few wires in the center of the design and firing when the lily root is dry. Working out from the center in small sections minimizes complications later. When creating figures, it is important to be aware of each part of the body because wires can shift easily and distort the appearance of the figure.

"Mermaid" necklace, 16", enamel 2.5" x 1.25". Cloisonné enamel on 26 ga fine silver (domed on hydraulic press). Fine silver granules for scales in tail, setting: fine and sterling silver seated bezel, freshwater pearls with enameled fine silver PMC beads and endcaps. *Photo by artist*

"Persephone" pendant, 1.5" sq, cloisonné on 26 ga fine silver, 24k gold wires, fused enamel grains, setting: 18k gold, 22k gold bezel, 3 faceted garnet bead drops. *Photo by artist*

After all the wires are fired in place, I apply Japanese gold or silver ginbari foils, which depend on the design. I prefer the Japanese ginbari foil because it is less delicate and can be textured before application. I place the foil on a clean piece of 100 grit emery cloth and gently press it with a clean brass brush. This simple process produces tiny holes that let air escape. I use either Klyr-Fyre or lily root to attach the foil to my piece. When dry, the piece is fired, and any wrinkles are pressed out with a steel burnisher.

When doing a multi-panel piece, I lay the pieces out as I want them to hang. Then I lay a strip of paper across them as a guide. If I am painting a shoreline, the design will line up throughout all the pieces. I work across all the panels at the same time, which minimizes color disasters from changes of kiln atmosphere from one day to another. After the general landscape, the sky is applied. I use about four to five layers of opal white for the clouds, Ninyomia P200, in watercolor-thin washes. I do not wash the white. It comes very finely ground, which makes it easy to apply in watercolor-thin washes. If this opal touches silver, it will turn yellow.

Once two or three layers of color are fired I can add gold. When using chain, I start with a length of 18k yellow gold chain and depletion gild it first. Depletion gilding is a process in which the gold is repeatedly heated and pickled. This process causes the copper in the outer layer to oxidize and be removed over successive heatings; it then leaves a skin of non-tarnishing gold.

A torch can be used, but I prefer using my 8" x 8" x 6" electric kiln at 1500°F for about 45 seconds or until the chain turns black: my kiln is already hot and heats evenly. I fire the chain on a clean mesh tray that is used only for the depletion gilding, since any enamel flecks will cause the chain to fuse to the mesh and leave bits of enamel on the gold. After each firing, the chain is put into a warm Sparex pickle for a few minutes, removed, rinsed in water, and brushed gently with a fiberglass brush. This process is repeated until the gold does not turn black when fired and stays a soft, buttery, matte yellow. The chain can then be fired into the enameled piece. I dip the chain into the lily root mixture and arrange it on a piece. Once fired, it will sink in a bit. Any cracks will disappear when the enamel is built up around and over the chain.

For water, I make small heaps of opaque white. I use two paintbrushes to press the damp enamel into wedges, let it dry and underfire so it will not lose any height. I continue with transparent blues and a sky blue to unite them all. To make rocks, I heap opaque enamels within the cells and underfire. Transparents and opals are added until I am happy both with color and the illusion of volume.

To impart a sense of depth for figures, I use opaques and opals to fill the first third of the depth of the cells. Surrounding the figures with mostly transparents pushes them forward in the image, and the eye perceives them as closer. Next I build up the human forms with a light opal that I mound in the cells. I remove any stray grains that are too close to the wire walls and fire the piece to orange peel. In the next layers, I use transparent browns to shade and give volume, again firing just to orange peel. Overfiring at this stage would cause the mounds to slump, and the illusion of curves would be lost.

With some enamel in all the cells, but not yet filled, it is time to remove any warping. I place each panel on top of the kiln for less temperature shock and gently press them with a pre-warmed cast iron grill press. The pressing is done in two stages rather than one overzealous squashing.

When the pieces are cool, I check for specks of dirt and straighten any crushed wires. Dirt can be drilled out with a small diamond-tipped burr. If wires are damaged, I move them back with pliers and tweezers. I use the shape that is the closest to the original shape of the wire. Such thin ribbons of fine gold require very little pressure to move them back. I fill in the remaining space in the cells with the N3 flux and fire. Then with diamond sponges and sticks, the pieces are ground under water. When the grinding is completed, I scrub each piece for at least five minutes with a fiberglass brush. The final firing is a flash firing for a smooth surface.

A piece such as this might take months to develop from a concept to reality. I like to play with images in my head to try to anticipate and avoid design or structural problems. The true joy is immersing myself for hours at a time in the creative process. ∎

"Leda and the Swan" necklace, 3"x 2", cloisonné on 26 ga fine silver, 24k gold wires, setting: sterling silver, 18k gold, 22k gold bezel, handmade sterling silver chain.

"Narcissus" (detail) necklace, 4 panels, 4" x 3", central panel 1.25" sq, Echo approx. 1" sq each, drop and earrings .66".

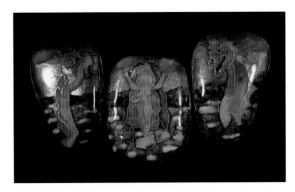

"Andromeda on the Rocks" triptych necklace, 5.25" x 2.25", cloisonné on 26 ga domed fine silver, 24k gold wires, silver foil, 224k gold granules, setting: 18k gold and 22k gold seated bezels.

Cloisonné on Fine Silver with 24k Wires

MERRY-LEE RAE

Merry-Lee Rae has been making jewelry since 1966 when she joined her father in the garage while he dabbled in metal arts. In 1976 she fell in love with enameling, and her life took a turn from hard-core academics to a relentless, obsessive pursuit of crafts. She feels privileged to have the support of her extraordinary family. She sees herself as a dedicated romantic. Her awards and achievements, the shows, the galleries, the private collections, and the lack of training are not as important to her as the people who are touched by her work.

Since 1976, I have been designing and making cloisonné jewelry, ranging from 1/2" to 2". My method of working is primarily the result of trial and error. After thousands of pieces and as many mistakes, I now have a narrow but proven approach. Each piece is started with a detailed drawing, including the plans for any goldsmithing and gemstones. My color test plates are used to select the enamels I will use. Progress notes are kept as a journal to note any unexpected variations from the firing. If there are any technical difficulties, my notes from other pieces often help to find the cause.

My kiln is a "Firemaster" with a 9"w x 6-1/2"h x 11"d chamber. To reduce drafts in the kiln, the peephole is covered with mica. The mica is held in place with masking tape. I rely on the pyrometer. I fire the enamel piece on a stainless steel trivet on a small firing rack. When firing, I wear welder's glasses to protect my eyes and a welder's glove and copper tongs to place the work in the kiln.

The shape is usually cut from 26 ga fine silver from the drawing and domed either with a Bonny Doon Hydraulic Press or by hand in a wooden dapping block. For most designs, I polish the front surface with white diamond tripoli, clean ultrasonically in non-sudsy ammonia and then brass brush with detergent under running water.

I use 80 mesh Thompson or Japanese leaded transparent enamels. The enamels are washed in small quantities with distilled water in a glass, stirring with plastic spoons or a glass stirringrod. The base coat on the front is Thompson's old flux for silver, #1209 or #757. When my supply is gone, I shall have to figure out something else. I spray the piece with a 1:2 solution of Klyr-Fyre and distilled water before sifting. This flux base coat is a very lightly sifted one that barely covers and is fired on a trivet at 1500°F for 70 seconds or until smooth. I try not to look in the kiln before the time is complete. Then on the back, a light to medium color of transparent blue is used because it holds up well and enables me to see variations in the thickness of the counter enamel. The back receives two to three heavy applications. The piece is then ready for the cloisons.

In general, my cloisonné wire is .003 x .060 24k yellow gold that I roll down from .005 x .050 wire. With tweezers and my fingers, I bend the wires, cut them and place them in position on the piece with uncut Klyr-Fyre. When it is dry, the piece is fired to embed the wires in the base coat of enamel. I carefully inspect each wire to be certain that all my partitions have remained in place and that each joint is tight so that one color will not bleed into an adjoining cell.

"Cayman Turtle" pendant, 2.5" w, cloisonné enamel on fine silver, 24k gold wires, frame: 18k gold, diamonds, purple sapphire.

"Mermaid Treasure" pendant, 1.5" w, cloisonné enamel on fine silver, 24k gold wires, frame: 18k gold, fresh water pearl.

"Snow Leopard" pendant, 2.5" w, cloisonné enamel on fine silver, 24k gold wires, frame: sterling, moonstone.

When the wire design is embedded in the flux base coat, the piece is placed on a Pyrex lid, which ensures a clean surface and allows me to rotate the piece without touching it. The lid also gives a slippery surface to slide the piece easily to the edge and transfer it to a trivet. I use plastic spoons to hold the washed enamels and I wet pack them with a fine brush and dental tools.

I achieve depth and subtle shading by firing many layers of enamel. Most pieces average from 10 to 20 firings. The majority of my shading is accomplished in the first five layers; any subsequent enameling helps to fine tune and bring all cells to the upper edge of the cloisonné wire.

When the enameling is complete, the finishing starts with lapidary equipment to grind the surface of the enamel. The equipment comes set up with a water hose and catch basin. The enamel piece is stuck to a dop stick to make it easy to hold against the wheel. To do this, I warm the enamel piece by placing it face down on top of my kiln and then I heat the dop wax in a small metal pot. Some warm wax is gathered on one end of a short wooden dowel. This wax end is placed on the back of the enamel piece. I now have a little handle to hold onto during the grinding. Be careful not to drip any wax on yourself or on anything that will go into the kiln. The front surface of the enamel piece should be free of any wax.

My lapidary grinder has an 8" expandable drum. The sanding belt I now use is a 40MIC microfinishing made by 3M, although for years I successfully used a 320 grit Carborundum belt. The equipment comes with a watering system because the process needs to be done very wet. The grinding is done carefully until all the wires are exposed and all glassy dips in the enamel have disappeared. Careful attention to the edges of the piece requires grinding a sloping edge that is evenly rounded.

I remove the dop stick by placing it in the freezer for a few minutes and letting the piece drop off in the warmth of my palm. Any wax residue is cleaned off with a knife. Holding the piece perpendicular to the sanding belt, remove any enamel from the edges. If you have an irregular edge, a large diamond bit in your Foredom handpiece will aid in this process. Remember to do all grinding wet. Next, the piece is vigorously cleaned with a glass brush under running water. Check that no speck of wax remains, as it will fire into the piece causing cloudiness and bubbles. I use clean, white, unscented paper towels to dry the work and then check that all the wires are exposed. If not, I regrind and glass brush. When the grinding is complete, the piece receives its final firing for a glossy and smooth surface.

If your wires extend to the edge of the piece, you may find a little lump at each wire edge, which could make a problem for burnishing a bezel smoothly over the edge of the enamel. I use my judgment at this point and often will return to the grinder, using a 15MIC sanding belt with about 600 grit, to refine the edge shape and then repeat the finishing steps and fire again.

A few tricks I have learned:

(When I say layer, I mean apply and fire the coat of enamel.)

• For a very light background, apply one to three thin layers in just the background, leaving the cells empty.

•For a dark background, I enamel everything but the background for one to three layers until the wirework is securely sealed to the base and then complete all the cells. The dark background can then be safely completed with several layers.

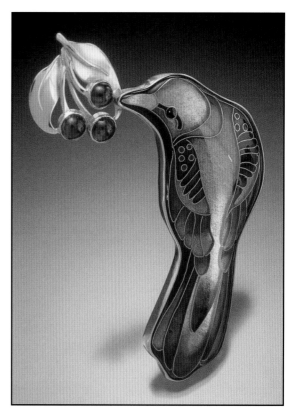

"Michele's Bluebird" pendant, 2" h, cloisonné enamel on fine silver, 24k gold wires, frame: 18k gold, Pyrope garnets.

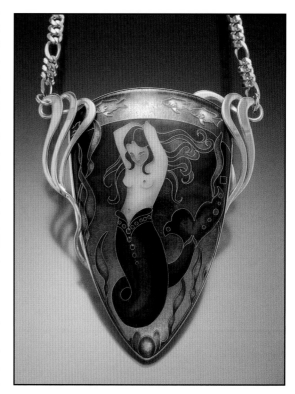

"Cayman Mermaid" pendant, 2" w, cloisonné enamel on fine silver, 24k gold wires, frame: 18k gold.

45

•To remove a speck, I carefully use a Foredom with a small diamond drill under water in a shallow bowl.

•I often use a design that has two long wires less than a millimeter apart. To prevent their being sucked together, tiny gold balls are placed between them in the first layer.

•To fill a cell with foil, I make a pattern by placing tracing paper over the embedded wires and do a gentle rubbing with the side of a pencil lead.

•For gold balls under the surface of the enamel, I wet pack a layer and push the balls into the enamel grains. Silver balls will sometimes cause stress cracks, as silver expands and contracts more than gold. To have gold balls on the surface, I complete the enameling and the grinding of the piece; then, using a diamond drill bit, I make depressions in the surface of the enamel for the balls, put them in place and re-fire to fuse them.

•If cracks appear in a finished piece made with Japanese leaded enamels on fine silver with 24k gold wires, the fault is probably with your counter enamel.

•Before the final firing in a larger piece, I warm my grandmother's antique iron on top of the kiln, fire the piece on mica, then quickly slide the piece and the mica onto a ceramic tile and gently weight it with the warm iron. It will hold its new shape for the final firing.

•To remove excess moisture before drying the enamel, I find that a folded square of unscented toilet paper that I pat gently over the enamel works well. I do not use the first few and the last few sheets of either the white Viva towels, which I prefer, or toilet paper because they have glue on them.

•To have a color deepen, I try to plan ahead so that I am adding colored transparents and not filling in with flux.

There is very much to be said for the merits of joyful experimentation and the fluidity of the creative process. There is no right or wrong way to approach this wonderful craft. I offer "my way" as a tool for you to find "your way." ∎

"Perfume Pendant" bottle, 6" h with detachable, wearable cloisonné enameled bead. Fabricated in sterling and 18k gold plique-à-jour enamel. Hand blown perfume bottle by Elaine Hyde.

LARGE CLOISONNÉ WALL PIECES ON COPPER

MARIAN SLEPIAN

Marian Slepian

Marian Slepian's site-specific installations for public spaces are seen nationally, and collectors have been accumulating her work for 35 years. Recently she has been making objects in fine silver, particularly for ritual use. She graduated from the Fashion Institute of Technology. Studying with Joseph Trippetti formed the basis for her enameling career, although she was self-taught prior to that. She currently teaches enameling at the Newark Art Museum.

I have worked in cloisonné on large-scale wall hangings and site-specific installations for public spaces. My work ranges from 8" x 10" to several feet. I have also created a walk-in outdoor sculpture that has withstood time and the elements. Recently, I have been making fine silver cloisonné objects, but this article deals only with my large cloisonné enamels on copper.

"She" wall piece, 14"w x 8"h, cloisonné enamel on copper, fine silver wires, *Photo by artist*

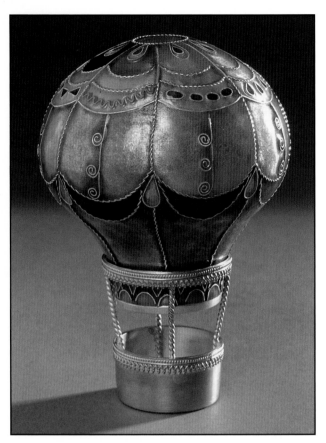

"Up and Away" sculpture, 5.5" h, cloisonné enamel on fine silver, fine silver wires, basket: sterling silver. *Photo by artist*

Communion Set, cloisonné on fine silver, fine silver wires, set in polished pewter. *Photo by artist*

I begin my work with a series of rough sketches on inexpensive newsprint I purchase by the bolt from a local newspaper. This paper is cheap, sturdy and large enough. I refine my sketches and then make a full size line drawing for the cloisonné wires. I do not color the drawings; I prefer to work the colors as I enamel.

My kiln is 18" x 18" x 12" on its own 220V line, so I have to make the enamel in sections no larger than 17". With this limitation in mind, I red line the drawing to indicate the cut lines for the metal. The cut pieces are my patterns for the 18 ga metal I cut with a metal shear and a nibbler. The nibbler cuts away a 1/4" strip of scrap from the metal. Copper is wasted this way, but for very large pieces a lot of time is saved. The nibbler is what I call my electric shear that is used by roofers and metal workers. It is heavy, and it takes two of us to move it around. If the enamel is one piece with straight sides, I sometimes bend the sides in a flange. I also use a small nibbler (available from Micro-Mark, 340 Snyder Ave. Berkeley Heights, NJ 07922) and an electric Dremel saw for fine cutting pieces that fit on the saw's table. Any etching is done after all the pieces are cut.

My kiln was custom-built; it has a pyrometer and three sets of independently controlled elements. Firing is kept to 1450°F to prevent the fine silver cloisonné wires from sinking into the copper and forming a eutectic, a metal alloy. The danger of this happening is increased by the size of the piece, so it is necessary to rotate the piece during the firing to distribute the heat more evenly. The areas closest to the elements will mature faster.

I use large amounts of silver cloisonné wire and, therefore, I can have it milled to my specifications. The wires that I order range from .005 to 16 ga, both round and flat. The wire is priced by the ounce, and there is a ten ounce minimum order of mixed sizes. I found that the 16 ga square wire could break loose in later firings.

Some pieces require more than ten firings. Unlike silver on silver, silver on copper has an inherent danger, for the more the piece is fired, the greater the risk of a eutectic and also severe warping. Warping makes mounting nearly impossible. To reduce warping, when the piece comes out of the kiln, and is still white-hot, I sandwich it between two 1/4" thick steel weights that are 18" square. I have developed muscles!

My leaded enamels are 80 mesh in the opaques and 100 mesh in the transparents. The back is counter enameled and a base coat of hard enamel is enameled on the front. These base coats are sifted. The drawing is transferred to the enamel with either carbon paper or fabric transfer paper, as both burn off cleanly. The wires are formed on the pencil drawing, bent with two jeweler's tweezers, cut, dipped in diluted Klyr-Fyre and set in place on the enameled base coat. Each piece is dusted with either opaque or transparent enamel after it has dried. It is then fired and cooled between the steel weights. The cloisonné wires do not distort from being weighted; they are pretty well anchored because I am careful to place them so that they touch the enamel. The very light dusting of enamel before the firing also helps to embed the wires in the enamel.

From this point on, all enameling is done by wet packing washed enamels. I use mostly opaques and then many layers of transparents for shading. Each subsequent firing requires strict attention to the kiln. A millisecond of too much heat can cause the sinking of some wires, which is a real problem to correct. You usually have to start that piece over again. For wet packing, I use a spatula to carry the wet enamels from their containers to the piece and a fine sable brush to pack them in place. I prefer multiple, thinly applied layers of enamel because I am ever cognizant of the possible eutectic; however, I do try to keep the number of firings as limited as possible. The entire piece is wet packed to the same level before each firing. ■

"Urban Fragments III" wall piece, 47"w x 22"h, cloisonné enamel on copper, fine silver wires. *Photo by artist*

"Who Shall Ride the Wind?" wall piece. 41"w x 23"h, cloisonné enamel on copper, fine silver wires. *Photo by artist*

The Burning Bush, a walk-in sculpture, 13.5' in circumference by 8.5' high, enamel on copper; a memorial to the 1.5 million Jewish children murdered in the Holocaust. Commissioned by Congregation Beth Israel in Owings Mills, Maryland.

Cloisonné on Steel

JOSEPH TRIPPETTI

"Musician," 16" x 16", silver cloisonné enamel on steel, fine silver wires.

"Carnival Midway" panel, 16" x 16", cloisonné enamel on steel, fine silver wires.

When Joseph Trippetti returned from the Army in 1946, he studied for three years at Philadelphia College of Art and the fourth year at Sheffield College of Arts and Crafts in England, where he majored in silversmithing. He has been enameling since the 1950s. For some years he taught enameling and painting before concentrating on commissions and gallery exhibitions. The medieval tapestries still influence his designs. His cloisonnés were on domed copper plaques before he turned to large steel tiles.

Design is my main interest. My method of enameling has remained about the same these many years. Originally, my work was mainly of cloisonné on domed copper plaques. I trained as a metalsmith. For the past 15 years I have been working on white pre-coated, flanged, steel plaques, ranging in size from 6" x 6" to 16" x 20". Using the pre-coated, steel tiles I do not have to be concerned with cleaning the metal and applying base coats. The fine silver, rectangular cloisonné wire I use is .010 x .035.

I have two Norman kilns, 15" x 15" x 9" and 27" x 24" x 15", on a 220V line. They were made for me with the specification to heat within 35 minutes and have fast recovery. Each kiln has a pyrometer, and I fire between 1250°F and 1500°F. My pyrometer has not been checked for years, and so my kiln temperature may be way off a standard, but it works for me. The floor of the kiln is protected with a Fiberfax blanket. Firebricks support the flanged piece in the kiln. I do use a timer, especially for the larger pieces, to remind me to look in the kiln after 3 to 4 minutes, at which time the piece is usually about orange peel stage. This method is adequate for all except the final firing. I usually eyeball it. Originally the kiln wires were exposed, but when a pitting problem developed, the wires were changed to be covered in the floor of the kiln.

I start with rough sketches in pencil and then translate the selected one to a full size ink drawing. Using carbon paper, like dressmaker carbon paper that leaves no residue, the pen drawing is transferred to the pre-coated steel plaque. To protect the drawing, I tape a sheet of glass to foam board and make a sandwich into which I slip the drawing. The pen drawing is used as a pattern to bend the cloisonné wires. I form the wires on top of the glass and then position each wire on the transferred design on the plaque.

The tool for bending the wires is one I designed by soldering a handle of the tweezers to one of the handles of straight bezel shears. The tweezers and my fingers are used to bend the wire; the short blades of the bezel shears cut the wire in place on the glass. My aim is to take the least complicated approach. The cloisonné wires are put in place on the plaque with uncut Klyr-Fyre. After the Klyr-Fyre has dried, the piece is fired. With my kiln at 1300°F, a 16" x 16" plaque is placed in the kiln and the timer set for 3 to 4 minutes. Subsequent firings are at around 1300°F to avoid over-firing the piece. Through all the firings, as with silver cloisonné wires on copper, over-firing can cause the wires to sink into the enamel.

I use primarily 80 mesh opaque, leaded, unwashed enamels, though I also have some 150 mesh enamels and some unleaded enamels that I use when I need those colors. To use them all in one piece, the unleaded enamel needs to be under the leaded enamel and not on top. The enamels, wet with water, are wet packed with a brush almost to the top of the wires, and then the piece is tapped to level out the enamel and fired. Before each firing, any opaque enamel on the wires is removed with a fine pointed brush. It usually takes about eight to ten applications of the enamel, tapping and firing for the fired enamel to reach almost the top of the wires.

The final firing, with just a thin sifting of either soft or medium flux over the whole piece, is a healthy firing with the kiln at 1500°F before inserting the plaque into the kiln. I do not wet the piece for the sifted coat. My sifters are made of 80 mesh screen bent into open boxes in square or rectangular shapes. The square ones are about 2-1/2" x 1/2" deep. I also have ones that I soft soldered together out of brass tubing.

I do not remove the veil of flux from the wires after the final firing. This coating protects the fine silver wires from discoloring. You need to be careful not to overfire this final firing in order to prevent the flux on the wires from balling up. For me, the most important stage in making each enamel is the pen drawing of my design.■

"Deer Disc" panel, 12" x 12", cloisonné enamel on steel, fine silver wires.

"Gardeners" panel, 12" x 12", cloisonné enamel on steel, fine silver wires.

"Bouquet" panel, 16" x 16", cloisonné enamel on steel, fine silver wires.

"Musicians" panel, 16" x 20", cloisonné enamel on steel, fine silver wires.

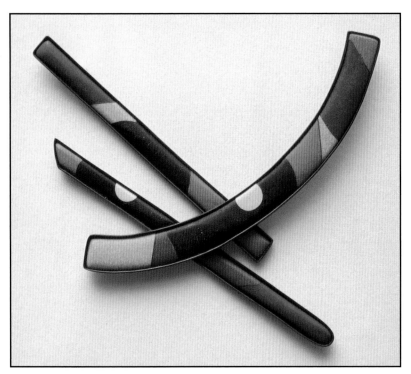

Brooch, 3.52" x 3.75", opaque matte enamel, sterling silver fabricated setting.

Brooch, 3.52" x 3.75", opaque matte enamel, sterling silver fabricated setting.

CLOISONNÉ OPAQUE ENAMEL JEWELRY

GINNY WHITNEY

Ginny Whitney, like the other professional enamelists in this book, has a curriculum vitae too lengthy to print here. She began enameling with Bob Kulicke in 1968, and in the years that followed had many metalsmithing teachers, whom she sought out when she needed a certain technique. Some of her enameling knowledge comes from her own experimenting. The patterns and colors of Russian Constructivist work influence her enamels. Color is the starting point and a dominant feature of all her work.

I make brooches and pendants in abstract designs. For cloisonné pieces I use either fine silver, 24k gold or milled binding wires. When the enameling is completed, the piece is hand stoned to a matte, smooth surface. I rarely do multiples. My kiln is a medium size, conventional electric one.

A piece starts with a rough idea of size and colors and then proceeds to detailed working drawings which show colors, cloisonné wires if used, shape of piece and sometimes placement of pin findings or location of cable if applicable. My work ranges from 1/2" to 7". I keep careful color notes of my work for reference instead of making enamel color samples.

The metals used for the enameling are either copper or fine silver, sawed out from a 24 ga sheet. The metal is cleaned with pumice, followed by a final wipe with denatured alcohol solvent, then rinsed and dried. I use leaded Thompson and Schauer 80 mesh vitreous enamels in hard, medium and soft fusing. Frequently, I mix several colors together to get the desired color. Each enamel is washed with tap water in the traditional method. I fill the working container with about a tablespoon of enamel, add a lot of water, stir and let the enamel settle, pour off the cloudy water and repeat until the water is clear. I wet pack the wet enamels with a fine brush.

The cloisonné wires of binding wire are made by milling down various gauges until I get the size line I want. This cloisonné wire provides the line that separates the colors. For the fine silver or gold cloisonné wire, I cut a narrow ribbon off a piece of bezel or sheet and proceed as above. I use the binding wire cloisons when I want a black line without any intricate curves because it is not as soft as the fine silver and fine gold.

My pieces are counter enameled with medium flux that is wet packed on the back, dried with tissue and then turned over onto a piece of mica. Cloisonné wires are placed on the top surface, and the wet opaque enamels are applied with a small brush or fine metal spatula. When the first application is complete, the piece, with underlying mica, is carefully placed on a stainless grid and dried slowly under an electric light. The drying can take from 15 to 20 minutes. The whole unit—enamel, mica, and grid—goes into a preheated 1500°F kiln. I fire between 1450°F and 1500°F. When the enamel is fused, the unit is removed and allowed to cool. Another layer of enamel is applied and fired. Then third and fourth applications are made to the front and flux is reapplied to the underside if needed.

The mica is pulled off the back when the enameling is completed and the piece is cool. The piece is stoned with a wet carborundum stone to even the surface, change the shiny, glassy surface to matte and produce a smooth surface. Another task of the stoning process is to slope the edge evenly like a cabochon.

I construct a bezel setting with fine silver and sterling or with 24k and 18k gold, with the appropriate findings for the brooch or pendant. ∎

Plique-à-jour: Russian Soldering Method

SANDELLE, *Sandra E. Bradshaw*

Sandra Bradshaw has been producing enamels since 1974. At age 18 she loved figure drawing; at 21 she discovered enameling and has continued the study of both since then in college and in private classes. In addition to plique-à-jour jewelry, she also creates small sculptures and watercolor paintings. Her teachers in plique-à-jour were her former business partner, John Ryan, and his original teacher, Valeri Timofeev.

Sandra Ellen Bradshaw in her studio. *Photo by Stephanie Kaehler*

A number of methods exist for plique-à-jour, a filigree technique that allows light to shine through the enamel, much like stained glass. With plique-à-jour, there is no backing for the vitreous enamel; instead a structure of metal is made with open spaces, called cells, for the enamels. I form a framework of fine silver square wire and within that framework I solder wires together in a design formed of cells. To make the framework as I do, you need to know how to use the jeweler's torch and how to solder. I use this Russian method of soldering primarily to make fine silver earrings. As the earrings dangle, the light shines through the enameled cells.

I start with rough sketches and then make an exact drawing. I keep a file of my drawings. If the earrings are asymmetrical, I also trace the drawing from the opposite direction. The cells are kept under 3/16" in width. The drawing is copied on tracing paper and trimmed with a 1/16" margin all around. The tracing paper drawing is the pattern on which I will glue the cut pieces of wire. I put this drawing on a 3" x 5" white index card and then put both on a 2" thick Styrofoam block. The drawing and card are held in place with stainless steel dressmaker straight pins, one in each corner.

I use 16-ga square, fine silver wire for the outer wire frame. The major lines of the design are 16 ga x 22 ga fine silver wire and the fill-in lines are 16 ga x 32 ga. You can use decorative wires if they are the same dimensions as the fill-in wires. I bend and cut the wires to the file card pattern with jeweler's pliers. Each wire piece, formed to match the pattern, is flattened on the steel block with a plastic hammer. The flattened segment is picked up with tweezers, dipped in Sobo white glue, and then placed on the tracing paper pattern. Be careful to keep the wires perpendicular to each other. Like the wires in cloisonné, they have to make a tight-butted joint.

When all the glued wires are in place and dry, I remove the pins and put the piece and the tracing paper on a 1/16" or 1/8" thick oxidized steel plate. The oxidization on the steel will prevent the solder from adhering to the steel plate. The plate is oxidized by being repeatedly fired in the kiln until it is black. Using .60 mm binding wire, bind the glued wires to the steel plate. First wind the binding wire horizontally 1/8" apart and then vertically also every 1/8" apart. When a piece of wire is holding every joining, the work is ready to be soldered.

The flux I use for soldering is a solution of half dry boric acid and half borax (Boraxo) mixed with warm water. About a half cup of each is put in a heat resistant glass-lidded container, like a small Pyrex casse-

"On Amber Wings" pendant and earring set, 3" x 2" x .05", plique-à-jour enamel on fine silver, baroque pearl, chalcedony.

"Desert Flower" bowl, 2.25" diameter by 1.5", plique-à-jour enamel on fine silver, setting: sterling silver.

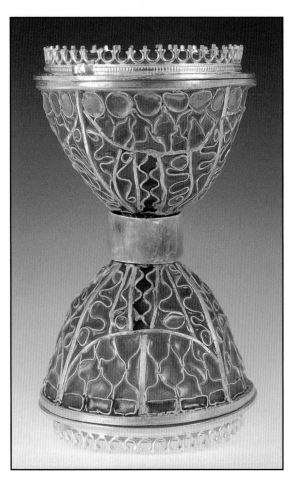

"Hour Glass Figure—Play with your Diamonds," 3.5" x 2" diameter, plique-à-jour enamel on fine silver, pixie dust, glass, and cubic zirconium.

"Squids" earrings, 2.5" x 1.25" x .05", plique-à-jour enamel on fine silver.

role with a wide opening, because hot enamels will be put into it. The water is added gradually while being stirred with a thin rubber spatula until the solution is the consistency of maple syrup. It should not crystallize when it is cool. If it does, add a little more warm water and let it cool. Eventually, it will not crystallize and that is just right. The flux solution is stored in a tightly covered glass jar so it will not evaporate. A hot plate or the top of a hot kiln can warm it when you are ready to use it.

The solder is 72/28 silver/copper that comes in a bar form. Thompson Enamel and T. B. Hagstoz and Sons, Inc. sell it. I make filings of the solder bar with a clean coarse file onto a piece of clean paper. Once you start to add the solder to the piece, you have to work quickly. The wired piece is dipped into the warm flux solution and placed vertically in a bird's nest (a wad of binding wire, loosely scrunched together) to prop up the piece. I sprinkle the solder filings with my fingers into all the joints. The flux is dried with a Rosebud torch tip, using a low oxygen bushy flame. The torch burns off the tracing paper also. Then the flame is increased, and the torch is moved up evenly from the bottom to the top as the solder flows. This solder turns black when it has flowed. All joints must be touching in order for the process to work. The piece is dropped into warm boric solution to cool and then the joints are checked. If there are any loose joints, I push them together with tweezers, dip the piece in the flux solution and reapply the solder filings with a brush. It usually takes about three soldering passes to complete the soldering. When all joints survive being tugged at with the tweezers, I remove the binding wire and drop the soldered piece into fresh warm pickle of Sparex2 solution, neutralize it in baking soda solution and rinse well. Remember to use copper tongs to remove the piece from the acid. The piece is rinsed well before the next three kiln firings.

The final step before enameling is to heat the piece in the kiln at 1430°F three times, for one minute each time, to bring fine silver to the surface. The first two times, the heated piece is put into water, then into an acid bath for five minutes, and then rinsed with clear water. The third time, the piece is not put in the acid bath but is either air cooled or cooled in water. Handle the piece as little as possible to avoid damaging it or contaminating it with oil from your hands.

My principle for plique-à-jour is short controlled firings that cause the enamel to just fuse to the metal cell walls. The firings must be short enough to avoid having the enamel pull from the center of the cells and run up the wires. It takes a well heat-saturated kiln to control the temperature drops when you open the kiln door. My kiln for plique-à-jour earrings is 10" x 8" x 6" on a 110 v line. To heat saturate the kiln, I fire it up to 1000°F and let it hold at that temperature for 20 to 30 minutes. Then I raise the temperature to 1450°F for firing. My kiln has a digital pyrometer, and I use a timer for firing. If you greatly over-fire the piece, then you have to grind off any unwanted enamels on the wires.

I purchase 80 mesh, leaded and unleaded enamels. I wash the leaded ones with distilled water, just the amount needed for one work session. The unleaded enamels are sifted to remove the fines. The sifted enamels are separated by +150 mesh and –150 mesh. In my experience there is no difference in clarity between 150 through 80 mesh and you need all those grain sizes for your mixture of gum, water and enamel to hold in the cell. I fill the cells with the +150 mesh unleaded and the 80 mesh, washed, leaded ones. To start enameling, I make a 1:5 solution of Klyr-Fyre and distilled water and cover each enamel with it, just enough to cover each grain of enamel.

To fill the cells, I use a dental tool that has a pointed spatula on each end, one large and one small. Hold the piece in mid-air between thumb and first finger and with the tool swish a gum/enamel mixture around the inside of each cell. I rub the side of the piece with the rough edge of the tool to vibrate the excess water to the surface and even out the grains of enamel in the bubble that was pulled across the cells. The excess water is blotted away with the point of a 3" diamond shaped piece of white blotter paper.

The cells are filled from the smallest to the largest and from the inside to the outside edge of the piece. Apply a thin coat. Some cells will take just a few fillings and firings while others may take eight to ten applications and firings. It is advisable to use opaque or less transparent enamels in cells that are most likely to crack at stress points. The stress points are the areas of transition: on a bowl, they are near the rim and on the shoulder; on earrings, they are in any sharp corners on the shoulder of curves. You learn by experience where the stresses are on each form.

Apply enamel to all the cells before the first firing. The softer enamels will fuse first, so concentrate on those. When making a pair of earrings, put them on trivets for the firing. Set your timer at 25 seconds when you place the piece in the kiln. If any enamel becomes slightly glossy, that is your firing time. If 25 seconds is inadequate, slightly cool the piece or pieces and then time for 28 to 30 seconds. Repeat the cooling and firing a little longer until the softest enamel reaches the slightly glossy stage. Any enamels that are sugary looking will need longer time after the others have a complete film. When the softest enamels have completed cells with no holes, I move to the higher firing enamel cells that are still sugary and add five seconds to the firing time until those cells start to become glossy. The filling and firing continues until there are no holes in any of the cells and all are glossy.

I check for holes with a magnifying glass under good light. The final firing is a little longer to give a true concave transparency. That final firing could be as much as 1 1/4 minutes, just until the enamel starts to climb the sides and thins slightly in the center.

To finish the metal, I first remove any enamel from the wires with the Airflex (332-710) wheels from Rio Grande. The metal must then be polished. To start polishing, I use the RoLoc system from Rio Grande. The diamond pads range from 180 to 800 grit for the flexible shaft. Sometimes the vibrations of the flex shaft will crack cells, so it is important to make certain the tool is well oiled to cut down on vibrations. Without a flexible shaft, you can start polishing the metal by hand with diamond papers. For my final polish I buff with tripoli and then with rouge.

Plique-á-jour is a challenging technique. This is my method now. Your job as an artist craftsman is to find creative applications for this ancient filigree technique.■

"Fall Leaves" earrings, 2" x 3" x .05", plique-à-jour enamel on fine silver, setting: sterling silver.

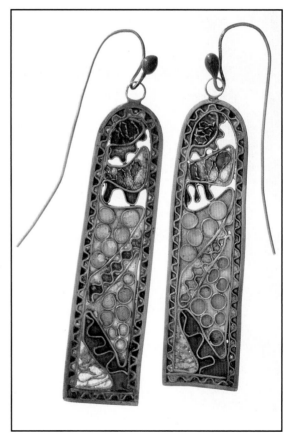

"Stained Windows" earrings, 1" x 3" x .05", plique-à-jour enamel on fine silver, setting: sterling silver.

CHAMPLEVÉ AND BASSE-TAILLE

"Introspection Instruments #1" hand mirrors approx. 8" x 3" x 1", champlevé/basse taille, enamel on copper, sterling silver, 24k gold foil, found objects. *Photo by Jack Zilker*

CHAMPLEVÉ: DUPONT'S IMAGON ULTRA FOR ACID RESIST

JAN HARRELL

Jan Harrell in the studio. *Photo by Taylor Harrell*

Jan Harrell has worked with enamels for thirty years. She earned a BFA at Texas Tech University in Lubbok, Texas, as a jewelry major and enameling minor. Since 1994, she has been the enameling instructor at The Museum of Fine Arts, Glassell School of Art, Houston, TX. Several galleries carry her jewelry, larger panels and sculpture. She has received numerous awards. Her work appears in Tim McCreight's The Metalsmith's Book of Boxes and Lockets *and* Color on Metal*.*

One of the first methods of enameling practiced by ancient craftsmen was champlevé, in which recessed areas of a metal piece are filled with enamel to create a glass inlay. In the 1800s, the shapes were cast in the metal to make lowered cells to receive molten enamel. In later years, the recesses were gouged into the metal with chasing and repoussé tools. Today, etching is used to create the depressions that are packed with wet enamels.

A variety of products, resists, have been used to keep areas of the design blocked as open areas are exposed to the effects of the etching medium. Etching with acid requires that the design be applied to the metal with a resist. While time honored resists like beeswax and asphaltum have their uses, new products make the etching of metal very much easier and the finished product more detailed. I use DuPont's ImagOn Ultra, a non-toxic, durable, dry film resist for deep etching. It is UV light sensitive and should be stored in a covered box. Keith Howard developed ImagOn for printmaking. He has articles on the Internet in addition to a book and a video.

The technique is based on exposing properly prepared plates to UV light sources and developing a resistant film coating that accommodates longer etching times and so gives a deep etch. The image produced, the black line work, is the recessed etched area that accepts the colored enamel. After the metal plates are laminated, care must be taken to protect them from light until the exposure of the image is complete. ImagOn Ultra differs from PNP and other resist materials in that the black drawing or photograph is the design that will be etched for the enamel. Suppliers sell it in rolls of 12" x 5' or 10' and 24" x 10'.

To begin the lamination process, prepare 16 ga copper by sanding in one direction with a 220/280 grit sandpaper. Clean the metal with an abrasive scouring powder like Comet, rinse thoroughly and dry. Swab the metal with rubbing alcohol. Do not touch the surface. This process is the same degreasing process used to prepare metal for enameling.

Next, cut the ImagOn Ultra film slightly larger than the copper plate. Handle it carefully as dimples in the film can be translated as imperfections later in the resist. The blue ImagOn Ultra comes sandwiched between two clear plastic layers. The inside layer has a soft feel like plastic lunch wrap and is matte, not shiny. The outside layer feels like stiff cellophane and is crisp and shiny (Mylar).

Mix a solution of 1 part 70% rubbing alcohol to 3 parts distilled water in a spray bottle. Place the clean copper plate on a cutting surface. Remove the inside matte peel-back layer from the film. An easy way to do this is to place duct tape sticky side up on a flat surface and press the matte side of the ImagOn Ultra down on it to aid in separating the layer. Pull the matte layer off.

Spritz the entire copper piece with a liberal amount of the alcohol solution. Place the ImagOn Ultra with the emulsion side down on the plate without making any folds and creases. Spritz the top of the film with the same solution. Then press a hard rubber squeegee out from the center using an X pattern. Repeat with increasing pressure until all the water and air bubbles are removed.

Trim the excess film from the edges of the plate. Next, using firm pressure, rub the top of the film with a dry cloth or sponge and ensure that all the edges are down. Bond the film to the plate by drying with a hair dryer or heat gun. Do not overheat the film in any one place. The drying depends on atmospheric conditions; it usually takes a few minutes. Film and plate MUST be completely dry before the exposure step.

The art work can be drawn directly on a transparency or acetate film with India ink or a permanent marker like a Sharpie. The images can also be computer generated and printed on a printer. If I am using a photograph, I can usually scan it so it is a black and white positive with no gray areas, which is all you need for photographs. To do this, the image might have to be adjusted on the computer in Photoshop or another program. In any case, the design should be a very dense black that does not allow any light to penetrate. The image that is produced, the black line work, will be the recessed etched area that is the area for the colored enamel.

I usually make a 2" x 4" sample piece to see whether the detail is good. You sometimes cannot tell about the detail with a smaller piece. The film can be exposed in the sunlight, and I develop the sample in the sun. (I live in Houston, TX.) The sample piece is also the timing piece because it takes the same time to develop as a 12"x 24" piece. To expose in the sun, make a sandwich of the metal and artwork acetate between two panes of glass and clamp it together with strong clothespin type clamps. Hold the package parallel to the sun. Exposure times will vary according to location, smog conditions and time of day, but the range is usually from one to two minutes.

Once you expose the sample outside, you use the developing solution for the full piece to see whether the timing is correct. If the image is underexposed, it is faint, and the pattern does not have great detail. If the piece is overexposed, the design lines get thicker, and detail is lost.

The results will be more consistent with a light box. Place the acetate of artwork down on a light box surface with the emulsion side of the printing facing the film side of the prepared metal. Close the cover and provide a secure seal with some pressure. Turn on the unit and expose. Timing varies according to the light box, and timing tests should be performed for each unit. My light source to develop the film was made from a black plastic tool box 26"w x 9"d x 10.5"h. The inside tray was replaced with a thick glass shelf. An under-counter light fixture with two 24" 20 watt black light bulbs was installed under the glass shelf. It cost about $80 to make.

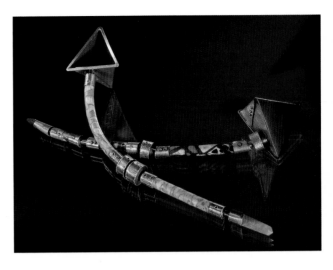

"Aphrodite's and Athena's Acolytes," candle snuffers, 12" x 3" x 1", champlevé, enamel on copper, sterling silver, 19k gold, quartz, river rock. *Photo by Jack Zilker*

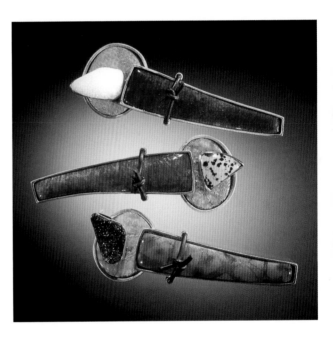

"Vertebrae Series #2" three brooches, 3.5" x 1" x 1"ea., champlevé, enamel on copper, sterling silver, rubber cord, 18k gold. *Photo by Jack* Zilker

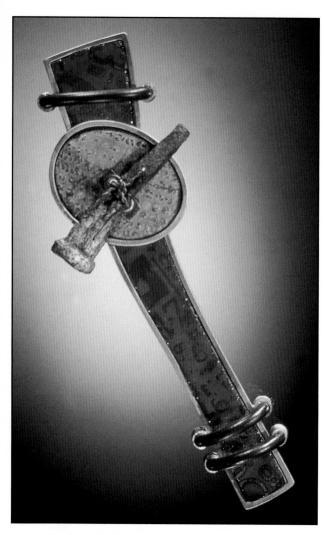

"Red Vertebrae" brooch, 4.5" x 1" x .5", champlevé enamel on copper, sterling silver, rubber cord, rusted nail, 24k gold. *Photo by Jack Zilker*

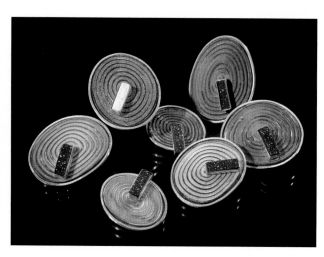

"Mazes" brooches, each approx. 3" x 1.5" x .5", champlevé, enamel on copper, sterling silver, drusy. *Photo by Jack Zilker*

After the plates have been exposed by either manner, remove the stiff Mylar coating by rubbing the edges with the palm of your hand or by pulling it off using the upturned duct tape. The plates are now ready to be developed. The ImagOn Ultra developer is a solution of 1 level tbsp. (10 g) of sodium carbonate (soda ash powder) to which about 1/4 cup of hot distilled water is added. Stir to dissolve. Make up 1 liter total volume with room temperature distilled water. The best working temperature is 650°F - 700°F. Pour the solution into a tray. The soda ash can be purchased at a pool supply store or as washing soda in the laundry section of a grocery store.

Set a timer for nine minutes. Place the plate into the developing solution, image side up. Put an opaque lid on the tray and start the timer. After nine minutes, remove the plate from the developer and run it under cold water. Using an ABSOLUTELY CLEAN sponge, rub the blue emulsion. The exposed image should be clearly visible. Be sure that there is no blue residue on the copper areas that are to be etched. The metal can be processed for a few more minutes if necessary. To stabilize the image, the final plate is rinsed with white vinegar. Fill a sprayer with white vinegar, spray the surface and lightly rub. Do a final rinse with cold water. The final image must be fully hardened by repeating the drying procedure. The plate is now ready to etch.

Duct tape is a good, durable backing for the exposed plate. Cover the back completely, and wrap 1/4" tape to the front of the plate to cover the edges. Place a large dollop of hot glue in the center of the duct tape backing. Then secure the back to a piece of 1" thick Styrofoam cut to the size of the metal plate. The plate with the Styrofoam becomes a "floating boat."

The ideal depth of the etch is about half the thickness of the gauge of the metal sheet. Ferric chloride is used for etching copper or brass, and ferric nitrate for fine silver and sterling silver. Radio Shack carries a mixed solution of ferric chloride labeled as PCB Board Etchant. These chemical salts provide a straight etch with little undercutting of the metal. They etch slowly, but the finished product is highly detailed and ideal for enameling. The copper takes about four hours; the silver takes longer.

The Styrofoam boat floats in a dish containing ferric chloride. A cheap aquarium pump is taped underneath the dish and provides agitation to help the scum that forms on the etching copper to be released. Check the boat about every half hour to be sure that no air bubbles have been trapped. The solution may have to be changed after a few hours if the etching slows too much or the solution appears saturated. When the depth of the etch is correct, neutralize the piece with baking soda. The resist is removed with a dilute ammonia solution. The acid turns green when it is spent. Check your local authorities for correct disposal of the spent ferric chloride.

The beautifully etched piece is ready to enamel. I saw out and form the shapes in a wood block or a stump with a rubber or rawhide mallet. The metal shapes are scrubbed with Comet cleanser, rinsed thoroughly and dried to be ready to be enameled. I use either 16 ga copper or 18 ga fine silver to make small color areas for jewelry and for sculptural objects.

I purchase the standard 80 mesh enamels and grade sift them with a 200 mesh sifter into a plus 200 and minus 200 mesh. I also wash my transparent enamels after grade sifting for the best clarity. I use my supply of Thompson's leaded 80 mesh enamels and fill in with some of the Japanese leaded enamels as my supplies dwindle. I like the Thompson unleaded fluxes, #2020 Clear for Silver and #2015 Golden Clear.

The first coat is usually an opaque or a flux. To remember that leaded enamels cannot be used under the unleaded, I say, "Keep the lead out." It is important to apply the base coat to cover the metal only enough so that no pits develop while not filling up too much of the etched area. Some experience is necessary to determine the number of applications and firings, which are somewhat restricted by the depth of the etch.

The first enamel coat is on the front in order to cover the etched metal: every time the piece firescales it loses some of the thickness of the etched metal that I worked so hard to get. This first coat also prevents a burnt out edge or at least so much of one. If I have some exposed areas of copper, I usually let it firescale. I sometimes clean the piece in cold Sparex2 between firings to eliminate any loose firescale that would get into the enamel between wet packings.

For detailed opaque wet packing, I use 325 mesh enamels with distilled water on flat pieces and a 1:4 Klyr-Fyre solution on domed pieces. Wick off the extra moisture with a Q-tip or absorbent paper towel and then vibrate the edge of the copper with a knurled stem tool to raise moisture and air pockets to the surface. For champlevé, I wet pack with a metal crow quill pen not a brush and direct the enamel into specific areas with the pen tip. A sprayer with the Klyr-Fyre solution is kept handy to remoisten the enamel, since it should not completely dry out while wet packing. I continue to add color and fire until the design is complete and the enamels are higher than the surrounding metal. Usually, my champlevé pieces take from five to eight packings and firings.

I have two electric kilns. One is a 6" x 6" 110 line Paragon and the other is a 17" x 17"x 8" 220 line Paragon. Digital controllers regulate the temperature: they are worth every penny! My firing range is between 1350°F for some soft, opaque enamels and 1650°F for some lead free fluxes. The firing range depends on the work and the desired effect.

When all the enameling is completed, the piece is stoned under running water with a 150 alundum stone, followed by a 100 and then a 200 diamond pad. It is scrubbed with a glass brush and rinsed well. The piece can be flash fired at this point or hand finished with additional finer grits of diamond pads or by using a lapidary machine.

The final appearance must then be chosen: should the exposed copper have a patina finish and what color patina; should the exposed copper be lacquered and kept a copper color or could a lightly sifted cover coat of soft or medium flux or a transparent enamel enhance the piece? These choices are critical to the success of the piece.

Enamels continue to seduce me despite more than 30 years of working in this medium. For the last 12 years, I have taught enameling. It has been rewarding for me to pass my experience and enthusiasm to new converts to this art. ∎

"Houston Concho/ Osaka Obi" belt, 48" x 2.5" x .5", champlevé enamel on copper, sterling silver, glass, found objects. *Photo by Jack Zilker*

CHAMPLEVÉ WITH FERRIC CHLORIDE

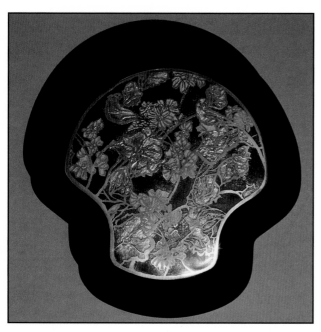

Pendant, 2.5"diameter, champlevé enamel on copper, gold and silver foil, 18k gold plated.

AUDREY KOMRAD

After graduating from New York University, Audrey Komrad started her artistic career as an oil painter, for which she received international recognition. Since 1970 she has concentrated on champlevé and cloisonné in enameling. She has taught extensively, her list of awards and exhibitions is lengthy, and her work is in several books. She was a founder of Enamel Guild South and presently serves on the executive board of The Enamelist Society.

Audrey B. Komrad

Champlevé (shahmp-luh-vay), meaning "raised plane" in French, is the process by which a design is cut, gouged or etched out of metal to create low areas for the enamel. After many layers of enamel have been fired, the piece is stoned and may be re-fired to restore a glossy finish. The un-enameled metal areas may be waxed, polished or plated. I cut, clean and etch many pieces at one time, but I fire each piece separately. Each piece requires approximately 8 to 10 firings.

Most of my work is either jewelry or small plaques. My jewelry usually ranges from 1-1/2" to 2-1/2" circles or interesting shapes. My plaques range from 4" to 7" circles and squares. I do a detailed drawing before I cut or select the copper piece. In my drawing, the lines for the resist, the un-enameled metal, are of various widths. None of the resist lines are narrower than 1/32". Although I do not work out my colors in the drawing, I determine where I want the darks and the lights, where the foils will go and if and where there will be cloisonné wires. I use Vidalon tracing paper to transfer the drawing to the copper after it has been annealed, cleaned and brushed.

The copper sheets I buy are 10 ga, 12 ga and 14 ga. I use a 1/0 or 0 blade in a jeweler's saw to cut the copper. The thicker the copper, the deeper the etch can be. A deeper etch allows for more layers of transparents over the opaques. I often dome the piece but not until the etching is completed. When I make a pendant, I have a "neck" as part of the piece. It will be turned back later to become a loop for the chain. The neck eliminates having to frame the enamel or to solder on a jump ring. I usually use 18 ga x 30 ga fine silver wire for the cloisons.

After the copper shapes are cut, the pieces are annealed. I place a number of pieces on a wire mesh trivet into the kiln at 1500°F. The copper is fired to a reddish glow and then plunged into cold water. Next is pickling in warm Sparex 2 *(follow the directions on the can)* to remove any firescale. Then comes the cleaning of the copper. I use Lea Compound C on a hard, stitched, muslin buff on my polishing machine. When the whole piece is buffed, it is rinsed very well with detergent and water. If you do not have a buffing wheel, a good copper cleaner will do; I like Copper-Glo. The piece is rinsed after buffing but additionally wiped with lighter fluid, rinsed again and dried. I then transfer my design to the metal with carbon paper as a pattern for the Weber's Liquid Etching Ground that I use as a resist. The resist prevents the metal under it from being etched. If in time it thickens, I thin it with mineral spirits. A border of resist is needed around all the edges. For a small piece of jewelry, paint a border about 1/8" and to 3/16" for a plaque. The border prevents

"Floral Fantasy" wall or table piece, 9.5"h x 13.5"w, sifted enamels on copper screening mounted on crystal Lucite.

the edge of the enamel from chipping or getting an underbite during the etching. Use a good, fine pointed brush. It is better to go over the lines several times with thin applications until there is no pink copper showing through than to apply the resist too thickly. Let the resist dry. Then, if you wish, you can sgraffito a design in the dried resist but remember that the resist lines should not be narrower than 1/32". If you are a beginner, leave these lines a little wider.

Having removed any resist from the areas that are to be enameled, I let the piece dry overnight before applying Bee's wax, paraffin or candle wax to the back and the edges of the piece. I melt the wax in an old frying pan, keeping the heat low to prevent the wax from smoking. With an old brush that I keep just for waxing, I paint on the wax, being careful not to disturb the resist on the front. The wax will cool and harden fast.

The set-up for the etching is a 14" x 9" x 2" deep Pyrex dish *(never metal)* for six to eight jewelry pieces. If I am etching fewer pieces, I use a smaller dish. A solution in a large container will etch faster than the same amount of solution in a small container because some surrounding air facilitates the etching. I place a row of triangular glass or plastic rods to support the pieces in the dish. I have had the best results with Ferric Chloride, even though nitric acid will etch the copper faster. Ferric Chloride etches by gravity so the pieces are placed **upside down** on the rods. They are supported only on the acid resist border.

The acid solution is poured slowly into the dish until the level of the liquid is just touching the under surface of the piece. That side has the design. The biting action of the solution takes place on the surface of the liquid. From time to time, I stir the solution with a pigeon feather, moving it gently over the surface to remove any bubbles. If you cannot find a feather, a wooden stick will do. Any bubbles trapped on the surface of the copper will retard the etching.

The ferric chloride is safer because it does not have dangerous fumes and produces no underbite, if used properly. I purchase a five-pound bottle of the purified lumps. The lumps keep better than either the liquid or the powder. Ferric chloride does not etch silver.

The solution I use is 13 oz *(avoirdupois)* ferric chloride lumps in two quarts warm water. Put the water in a plastic bucket and add the ferric lumps. **Caution: Aways add acid to water.** I wear heavy rubber gloves to break up the lumps in the water until they are in solution. Although this acid will not burn your hands like nitric does, it can yellow the skin.

I check the etch occasionally. I have a bowl of water next to the etching dish. Wearing the special rubber gloves, I remove the piece and place it immediately in the bowl of water. Do not allow it to be exposed to the air or a green powder will form, which happens when ferric solution is exposed to the air. **Important:** if this green powder is heated in a kiln, it forms chlorine gas, which is extremely dangerous to inhale.

While the piece is under the water, the depth of the etch is checked with a stainless metal tool. If you see a bright spot, rub this spot with the tool. These small, round, bright spots mean that an air bubble has been trapped there. The tool will clean the places on the metal where it may not be etching properly. If any of the acid resist lines are lifting up, the piece is placed in a bowl of water with a little ammonia for a few minutes, then into a bowl of clear water. The piece is dried and resist applied where needed. The resist is allowed to dry and then replaced in the ferric solution. With the solution I use, the etch takes about eight to ten hours. I usually start it early in the morning to be able to check it throughout the day. I etch several pieces at a time, but if I etch only one piece, the time will be shorter. It is always better to do a long, slow etch to avoid underbite. Underbite is when the sidewall of the etched area becomes uneven.

When the solution turns murky green it is exhausted. Remove the pieces and place them in a bowl of water while you replace the solution.

"Color Me Beautiful" pendant, 2.5" diameter, champlevé enamel on copper, gold and silver foil paillons, 18k gold plated.

"Star Gazer" wall piece, 7" x 7", champlevé enamel on copper, 18k gold plated.

To dispose of the acid solution, I neutralize it with a little ammonia in another container and pour it into a small hole I dig in the corner of my backyard.

To do multiple bites having different levels, I make the first bite about one third of the total thickness of the metal. The second and third bites together should go down no more than two thirds of the total thickness of the metal if you are using 10 ga or 12 ga. If the copper is 14 ga, do not etch down more than halfway for total bites because then the metal will be too weak. If I want other shallower etching areas, I scratch away some of the resist when the etching is about halfway done. You can repeat this after another few hours. I use a piece of 18 ga x 30 ga flat cloisonné wire to test the depth of the etch. The multiple bites in foil areas will make the foil appear to float.

After the etching is completed, wash each piece in water and soak in ammonia for 10 minutes, then rinse very well. I remove the etching ground with either paint thinner or turpentine. The wax is removed by heating the piece gently in another old, shallow frying pan. Some of the etching ground can be removed with this method also. When all the resist and wax have been removed, the metal is scrubbed with an old toothbrush and a mixture of hot water, detergent and a little ammonia. After being rinsed very well with running water, the work is pickled in warm Sparex2 solution for 15 minutes, washed well and dried. If I want to dome a piece, I anneal it as before and then use a sandbag followed by a metal mushroom stake and a rawhide mallet.

Next I drill any holes that will be needed. If I have left a neck for a pendant, I turn it back into a loop for a chain. Finally, the piece is rubbed with an 8" brass brush until the metal is shiny. It is then rinsed well and ready to be enameled. If you are not enameling immediately, wrap the polished metal in paper towels.

I use 80 mesh leaded, mostly medium fusing enamels and some soft fusing enamels. I wash small amounts of enamel, usually enough for a day's work. The enamels are washed about 10 times with tap water and then twice with distilled water. For wet inlaying, I use a 00 or 000 pointed sable brush. With distilled water, I wet inlay all colors thinly to avoid trapping air that will cause pits. In the deeper etched areas, I often use a light opaque under the transparents. To keep the color bright, I use a lighter transparent under a deeper transparent. If I need to lighten a transparent color, I mix flux with it. If you use opaques alone, put them into the narrow areas and use transparents for the larger areas where more light can enter. I generally fire at a little under 1450°F to orange peel stage, except for the final firing which is to maturity.

I fire the base coats in two firings, the back first with the front protected with Scalex, a firescale inhibitor. Keep a sepa-

"Creation" wall piece, 11.5 h x 10.5" w, silver cloisonné enamel on copper, gold and silver foil mounted on multiple layers of crystal Lucite.

rate brush for the Scalex. Using a color compatible with the front color scheme, I do three light siftings on the back using Klyr-Fyre diluted 1/3 to 2/3 water. When dry, the piece is fired to orange peel and cooled. The Scalex is washed off. If some of it remains, use a stiff toothbrush, a glass brush or a dental tool to remove the last of it. The piece is pickled in warm Sparex2 solution for about ten minutes, rinsed and dried. Then I brass brush the front of the piece until shiny and rinse again. On the front, I do two very light siftings of medium fusing flux, spraying between siftings with the Klyr-Fyre solution. A little of the pink copper may show through. Fluxing over the entire front of the piece prevents the buildup of a heavy coat of firescale in the subsequent firings. This coat of flux will be stoned off later where the copper is to be bare.

The deepest and largest areas that are to have foil or cloisonné wires are set up first. The deepest areas to have only foil receive one or two additional layers and give the foil a floating effect. I cut the foil between tracing paper with small, sharp, pointed scissors. I like the foils cut small or finely minced. Paper punches work well, too. To position the foil, I use a small pointed brush with distilled water that has a drop of Klyr-Fyre. Firescale on the walls of a cell can destroy the edges of silver foil. To prevent that, a thin coat of either a light color, medium fusing, washed opaque or flux is wet packed on the cell walls with water and fired before setting in the foil. Firescale does not affect gold foil. I have found that repeated firings of warm colors over silver foil sometimes produce undesirable changes in the color of the transparents. Finally, all the areas are enameled in thin-fired layers until the enamel in the cells are slightly higher than the bare copper. The final firing is to maturity.

You have a choice for the enamel finish: matte, semi-gloss or high-gloss. A matte finish is my favorite. It is a long process. I use alundum stones in 150 and 220 grits. They are about 6" long and 1" thick. The 150 grit comes in a smaller size and is good for small, hard to reach areas. Under running water, I start with the 150 grit and stone in all directions. When the unenameled copper areas are almost clean, I move to a 220 grit or the fine side of a Carborundum stone. During the stoning, you need to check for low spots or pits in the enamel. If there are any, you need to clean the piece before adding enamel. I scrub the surface with a glass brush, soak the piece in a solution of detergent, ammonia and water for a few minutes and rinse thoroughly before I reapply and fire the enamel where needed. Then I continue the stoning process until the entire surface is smooth and dull. There will be a lot of scratches at this point. Next come wet and dry emery papers to remove the scratches. I use 220, 320, 400, and 600 grits. I cut them into 4" and 6" squares that I use over a block of wood. I start with 220 grit then go to the others, rubbing in one direction over the piece and then in the opposite direction. Each step removes the previous scratches. It is best not to rush through the stoning.

When all the scratches are out, I wash the piece with detergent, ammonia and water again, rinse and dry. At this point, I use a polishing machine. My machine has two spindles, 1/2 hp, 3450 rpm with a sealed motor and filter. For polishing, I first use a bobbing compound on a hard-sewn muslin buff; then I wash and dry the piece and go to a hard-sewn felt buffing wheel with tripoli. The next step is to polish with a soft felt wheel and red rouge followed by a soft chamois wheel. These polishing steps can also be done with a flexible shaft.

After the piece is polished, I wrap it in a soft cloth until I send it to be gold plated. The plater applies a coat of nickel before the gold plating which helps the gold plating last longer. The appearance of the final piece depends on how good a job you did with the stoning and the buffing of the metal. Take your time and you will be rewarded with a beautiful surface and a sensuous enamel. ■

"Spirit Dance" sculpture, 8 sided, each 11"h x 3.5" x 1.5" Silver cloisonné enamel on copper, gold and silver foil, pewter. Mounted on two free-standing, tapered, Lucite columns.

"Royal Ibex" brooch/pendant, 2"x 1" Basse-taille enamel, cast fine silver and 24k vermeil. *Photo by Ralph Gabriner*

BASSE-TAILLE: DIE STRUCK FINE SILVER

MICHELE RANEY

Michelle Raney

Michele Raney knew when she touched metal in her high school jewelry-making class it would be her life's work. She studied metal arts at San Jose State University, goldsmithing at Revere Academy and engraving and enameling at Sir John Cass Polytechnic in London. Her technique is an innovative version of the centuries-old French enameling method, basse-taille. Her designs are inspired by nature.

When I studied basse-taille enameling in London, I hand carved each piece in silver. After I returned to the United States and began to design a line of jewelry, I realized the need for an efficient method of production. This need lead me to die striking and then to electrical discharge machining (EDM) to make dies.

Die struck metal, like hand wrought metal, is more homogenous than cast metal and frequently stronger; cast metal can often be porous. Die struck metal is also ideal for enameling because it requires less clean up and maintains detail. The EDM process uses a graphite positive to make impression dies in an acid etching bath. I find graphite is an ideal material to carve. I use the same jewelry tools as if I were carving out a piece of silver: gravers, burs, files, sandpaper, and so forth. I buy the 1" x 3" x 6" graphite block EDM poco3, which I cut into various sizes to fit my designs. That grade is good for detail work. Designs for die striking cannot have any undercutting in the relief because the metal being formed by pressure must flow into the die without tearing or folding.

The images in my designs are usually related to nature. The design incorporates a frame as a built in setting for the enamel work. I start with a simple line drawing on vellum, which is transparent and can withstand lots of erasing. To transfer my design, I coat the graphite piece with a thin layer of white correction fluid such as Liquid Paper. A reverse drawing is traced on the backside of the vellum sheet with a soft lead pencil. The drawing right side up is then placed on the piece of graphite coated with the white correction fluid and traced with either a hard lead pencil or a blunt scribe. Then I scribe the design into the graphite block with a sharp tool.

The graphite piece is placed in a graver's ball for maneuverability while carving. Carving by hand allows me to remove material a little at a time. I spend hours, often days, detailing the carving. When the carving is completed, I send the piece to Harper Mfg., a shop that specializes in EDM die and stamp making. I have worked with Tim Harper from the start. With both of our skills, we have come to know what it takes to make an exceptional die from a single piece of hand-carved graphite. The steel for the dies is oil hardened O1. The scale of hardness that is ideal for my dies is 58 - 59 on the Rockwell scale. To achieve the ideal resistance to shock during stamping, the steel is first heat treated to a specific hardness, then tempered to a specific toughness. If the steel is too hard, it is brittle and will crack; if the steel is too soft, it will distort the design. I leave it up to the professionals to heat treat my dies.

In the coining process, I stamp out my pieces from 12 ga fine silver sheet blanks, annealed dead soft. The blanks are cut close to the size and shape of the design with a bench shear (Heinrich 3" cutting blade). If the blank is too large it tends to spread outward and does not fill the cavity completely. Too small a blank will not fill in the cavity. The size, detail and depth of the design determines the pressure range for stamping: 20 – 150

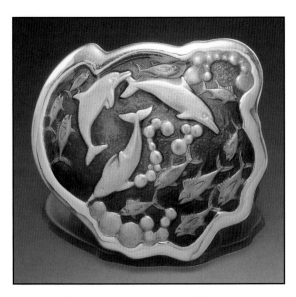

"Sea Turtle" brooch/pendant 1.25" Basse-taille enamel, graphite carving, die struck, cast fine silver. *Photo by Ralph Gabriner*

"Dolphin Tuna" brooch/pendant 1.875" Basse-taille enamel, graphite carving, die struck, cast fine silver. *Photo by Ralph Gabriner*

tons. My press is electric hydraulic and very slow moving, which makes it excellent for control. My pieces range from .5" to 2.125" diameter and a thickness of 14 ga to 16 ga. There are companies that will do the die striking.

After the coining process, I dome the coin with a wooden dapping punch in a wood dapping block and then polish the edges and back side with a combination of white diamond and tripoli compound on a 3", 42 ply loose-weave, coarse muslin buff for fast cutting. I rarely polish the front surface and usually prefer to leave a satin finish. The polish compounds are cleaned off with straight ammonia in my ultrasonic cleaner.

The findings are soldered in place with Batterns flux and hard silver solder. I use an acetylene gas with an air drawn Goss torch for soldering. A hot Sparex pickle bath removes the flux. Then the piece is put into the ultrasonic for a little more cleaning with a soft toothbrush and dish soap. For a slight shine, I use a soft brass brush and soap or for a dull finish, I use fine pumice powder with a soft brush and soap. I find that a highly polished finish is too reflective and harsh for my designs.

My kiln is a Neycraft 110-V stepless/manual dial, built in pyrometer, top vent. A molded ceramic fiber heating muffle protects the heating elements. It is well insulated and holds heat evenly. The chamber is 9"w x 9"d x 6-1/2"h. I fire all pieces face up; consequently my kiln is very clean. I wear didimium safety glasses with a light lens that makes it easy to see the work. A leather gardening glove protects my hand. The steel trivet and fork system is made by Frog Hollow Studio.

To start enameling, each piece is fired for one minute at 1500°F. This firing helps to prevent surface tension when wet-packed enamel is applied. The firing leaves a silvery white finish on the metal. I use Blythe, Cristallerie and Ninomiya transparent leaded enamels; they are all compatible. I prefer to enamel on fine silver, which does not firescale, because cleaning between firings is unnecessary and the transparent colors fire true and brilliant. Some colors, however, such as some reds and oranges do not work well directly on silver.

I grind and wash the enamels as I use them. Fine grains are needed to achieve the detail in all of my pieces. With tap water and the mortar I wash out the very fine particles, then place each color in a plastic spoon. I do a final rinse of the enamel with distilled water in each spoon. I wet pack all the layers of enamel with a very fine sable brush. The background and darkest shading colors are applied first. After drying on top of the kiln, the piece is fired for two to four minutes at 1500°F to full maturity. No counter enameling is necessary because of the thick gauge of the metal.

To achieve depth and shading, I fire enamels from dark to light. Simple designs require only two firings, more complicated designs up to eight firings. I keep a journal of what works and what doesn't. After the last firing, I have only the metal to clean up: bits of enamel always end up fused to the framework. I use my flex shaft with a light touch and a High Flex Blue polishing wheel from Rio Grande to remove the unwanted enamel on the frame. White diamond and a muslin buffing wheel give a final polish. Some of my pieces are embellished with a 24k gold plating.

I use an acid base plating solution from Rio Grande and a Procraft Electro-Plater. I place the solution in a 50 ml Pyrex beaker. For best results, I heat the solution to 110°F by placing the beaker in a pan of hot water. I hook my piece to the (-) wire and the 24k gold sheet (30 ga, 1" x 2") anode to the (+) wire. Each wire is placed into the solution without touching the other. I leave the solution in the water bath to keep warm during plating for best results. To keep the enamel from cracking, I plate at 2 volts maximum for two minutes. I can achieve only a thin plating, so I leave the gold accents in the recessed areas of my designs.

I have been carving and enameling for over twenty years, and my work constantly evolves. I could not achieve my work alone. Family, friends, and customers give me great support. My interaction with fellow artists and technicians has been the most important avenue for my achievements. My journey is challenging; it is a fun journey and it is a journey meant to be shared. ∎

"Polar Bear" brooch/pendant 1.25" Basse-taille enamel, graphite carving, die struck, cast fine silver. *Photo by Ralph Gabriner*

"Golden Eagle" brooch/pendant 1.25" Basse-taille enamel, graphite carving, die struck, cast fine silver. *Photo by Ralph Gabriner*

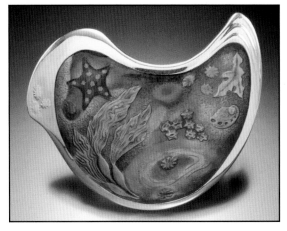

"Tide Pool" brooch/pendant 2.125"x1.5" Basse-taille enamel, graphite carving, die struck, cast fine silver. *Photo by Ralph Gabriner*

Liquid Flux as Etching Resist Basse-taille

PHYLLIS WALLEN

Editor's note: Phyllis described her technique for me more than twenty years ago. Phyllis died September 28, 2000. Her method of working is still current, and I am glad to be able to include it in this book. The liquid flux that Phyllis used as her etching resist was similar to the opalescent crackle Doris Hall used that predated the liquid flux.

In basse-taille (bahs ty), the metal has a design or a texture partially cut into its surface before being completely covered with transparent enamels, which allows the design or texture to be seen. Engraving and etching are among the many methods for cutting the metal. A resist is first painted on the areas that are not to be etched before the metal piece is placed in an acid bath. Among the materials that can be used for the resist are Klyr-Koat liquid flux, Thompson's medium fusing liquid flux and crackle. This description uses the Klyr-Koat liquid flux as the resist and nitric acid for the etching bath. After coating the metal with the liquid flux, the dried coat is sgraffitoed to expose the metal that will be etched. After the sgraffitoing, the liquid flux coat is fired. This fired coat of enamel is the resist for the etching.

The liquid flux is applied by dipping a cleaned 18 ga copper plate or bowl in a shallow bowl with the liquid flux. The liquid flux is first tested with a spoon to determine whether it is the right consistency. You can tell by the way it coats the spoon. If the flux is too thick, distilled water is added one drop at a time. If it is too thin, then it has to be set aside until some of the water evaporates. Some of the binder will be lost if you spill out the excess water. The material needs to be stirred often while it is being used because it settles to the bottom of the container. With experience you will learn the right thickness for this base coat. If it is too thick, then when it dries you will not be able to draw any fine lines in it.

After you dip the copper piece in the liquid flux, shake off the excess into the container and place the piece upside down on a hammock or trivet until the coating on the front and the back dries. The dry, unfired liquid flux is very sensitive and will show every touch or water mark. If you make a mistake in your drawing, remove all of it and dip the piece again. Patches will show when the piece is fired. The design or texture to be etched is sgraffitoed with any kind of a pointed tool; the finer the point, the thinner the etched line will be.

To prevent the dry liquid flux on the back of the piece from rubbing off while you draw on the front, place the piece on a slick surface, e.g., a piece of plastic or a shiny magazine cover. There will often be a thick accumulation of dried liquid flux around the face edge of the piece. This edge should be shaved down gently and gradually to almost the same thickness as the rest of the coat. You can use a 1/4" diameter dowel piece to thin down the dried liquid flux or any smooth tool. When the drawing is completed, the loose dry flux is tapped off and the piece is fired to maturity in the kiln at 1500°F. You will see your drawing in an oxidized line or pattern on the front and a layer of fired flux on the back when the piece is removed from the kiln. If there are any unwanted small pit marks in the flux, cover them with clear lacquer nail polish before placing the piece in the acid bath.

Make an etching bath solution of one part nitric acid to three parts water. **Never pour water into the acid; always pour acid into the water to dilute the acid.** Pour the water into a photographer's rubber tray and then gently pour in the nitric acid. You can also put the bath in a Pyrex container and then place it on an electric warming tray set on low. The nitric solution etches faster when it is warm, but do not let the solution get hot or it will melt any lacquer or cause the enamel to move. A 6" copper plate will etch in from 30 minutes to 1 hour. On a cold day in a cool solution, the etching can take 2 to 3 hours. Most writers say you get better etch with a slower etch, but Phyllis said she got a better line with a faster etch.

The depth of the cut in the metal, called the "bite" in etching, should be limited to one-third to one-half the thickness of the metal. Most enamelists use 14 ga to 16 ga for a deep bite and 18-ga for a shallow bite. Phyllis used 18 ga for her basse-taille plates and bowls.

When the etching process is complete, the piece is rinsed thoroughly, scrubbed with a brush under running water, and then dried. The oxidized lines are burnished with a glass fiber brush to brighten the copper so that they will be a golden flux color. For the first application of transparent, 80 mesh, leaded enamel, either Thompson's #333 hard flux or #728 Amber is used. The enamel has been washed and dried. It is sifted on dry and pressed into the etched areas. Care must be taken to pack the enamel well along the ridges of the etched lines.

If this first fired coat of the 80 mesh does not cover completely, the piece again is placed in the acid solution to remove the firescale and is rinsed well before another coat of that same transparent enamel is applied and fired. If the edge of the piece is ragged from being eaten away in the acid solution, it is ground or filed smooth at this point. Color transparents are applied and fired in thin layers; the firescale is removed from the edge of the piece between firings.

Generally, each coat is fired to maturity. The counter enamel is applied immediately after the first transparent firing on the front. If there are blemishes on the back, then the counter enamel needs to be a mixture of transparents and opaques to cover them. The counter enamel usually needs an additional coat as a last firing for the piece. Thick coats of transparent color may produce a cloudy final coat instead of the clear, brilliant transparent enamel that is desired. Make the last firing as fast and as high as you dare in order to add brilliancy to the transparent enamels. The edge of the piece is filed and then polished to a smooth finish to complete it.■

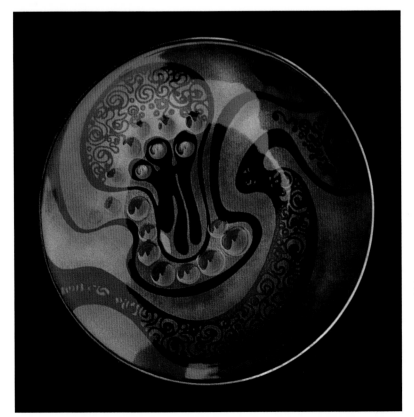

Plate, 7" diameter, copper, etched, basse-taille, enamel, circa 1974.

ENAMELING ON PRECIOUS METALS

ENAMELING ON STERLING SILVER

P. ALEXA FOLEY, *M.A.*

P. Alexa Foley began enameling in 1971. She studied with Joanna Stone, and then persevered with experimenting until she developed her method for enameling on sterling. She spent many years as a successful enamelist, creating cloisonné, champlevé and plique-à-jour enamels. As a realtor with a GRI, she now resides in Maui, HI where she does some enameling and painting. In June 2001, Alexa received her M.A. in Cultural Anthropology and Transformative Learning.

I enamel on various metals making a variety of pieces from post earrings to 5" x 8" framed enamels. The enameling techniques I use for sterling silver are cloisonné, champlevé and plique-à-jour. When I began enameling in 1971, it was accepted that true transparent enamels could not be used on sterling silver. After studying with Joanna Stone, I persevered with my experimentation until I developed a procedure for enameling on sterling silver. Every step is important.

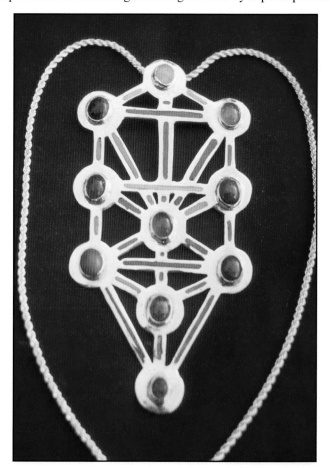

"Tree of Life," meditation necklace, 3" x 6", champlevé, sterling, fine silver, opaque enamel.

My sterling enamel jewelry pieces are made with a narrow sterling frame that I cut out with a jeweler's saw blade. The same pattern is used to cut out both the bottom piece and the frame. The frame is sweat soldered to the face of the sterling piece that is to be enameled. This method eliminates the need for making a bezel setting for the enamel. I also use this method for combining champlevé with the cloisonné. Instead of making just a frame, I saw out two full size pieces and then saw out cells in which I will enamel and place cloisonné wires in either some or all of the cells.

The sterling silver needs to be accurately alloyed of no less than 92/100 silver content. Reputable refiners can do this. I specify that the sterling is to be enameled and, if possible, the sheets are to be annealed. When I use 24 ga for the enameled piece, I select 18 ga for the frame. With an enamel piece of 18 ga, I use 20 ga to 22 ga for the frame. The 18 ga base needs only one counter enamel coat. Since the thinner gauge will warp easily, it needs about three coats of counter.

I often do an exact drawing using colored pencils. From the drawing, I make templates of 20 ga to 22 ga copper as patterns for the enamel piece and the frame. I use the jeweler's saw to cut these templates; then I scribe their shapes on the sterling sheet and saw them out. If the sterling is not annealed, I coat both sides of the two pieces with soldering flux (borax and water, or Battern's) and anneal them either with a torch or in the kiln. The flux and oxide are removed in a fresh hot pickle of 1:1 nitric solution.

Caution! Remember to add acid to water. Never add water to acid. When using acid, it is advisable to wear proper gloves, long sleeves, face mask and safety goggles. Sparex2 may be substituted for the nitric solution. If the sterling sheet came annealed, then it only needs to soak briefly in the fresh acid solution to remove all oils. To neutralize the pickle, I add water to the solution followed by the very slow addition of small quantities of baking soda. I wash the pieces thoroughly with fresh water.

Next, I borax-flux the piece and the frame and then secure them together with binding wire to be sweat soldered with IT silver solder or tested hard silver solder if I am unable to get the IT solder. There should be no solder on the area of the metal to be enameled. I bright dip the soldered unit to remove oxides. A freshly made 1:1 nitric acid solution at room temperature is needed for the bright dip. Sparex-2 also works well. Bright dipping is the most important step in the preparation of sterling for transparent enamels. The bright dip eats away a layer of sterling silver oxide, along with a layer of sterling silver. The trick is to know when to remove the piece from the acid before irreparable damage is done by extreme acid pitting. This depletion gilding is learned with experience. The soldered piece is placed gently in the acid bath. I use a Pyrex dish with a cover for the acid. Bubbles that rise to the surface of the metal are removed periodically by brushing across the piece with a long feather. When the piece is very dark, uniform gray in most places (it will be darkest around the solder seam), it is removed from the bright dip bath with wooden or plastic tongs, rinsed, neutralized with baking soda, rinsed and cleaned with a glass brush.

I put on rubber gloves to protect my skin from the minute glass threads that break off and can stick tenaciously and painfully. I rub the metal lightly with a glass brush in a circular motion under running water to minimize breakage of the glass brush. The glass brush removes the black residue of oxide that the acid brought to the surface.

At this point, the piece may be domed evenly and slightly with a rawhide mallet by striking the face of the piece that is held over a steel stake. The findings are soldered on the back with as little IT solder as possible, the firescale removed again and the piece glass-brushed as before. The piece is now ready to be enameled.

I use 80 mesh, transparent, unleaded enamels. I also purchase these enamels in lump form that I grind with a mortar and pestle when I am ready to use them. I prefer grinding my own. I wash the enamels, about a tablespoon of each ground color, with distilled water until the water is clear. To make the enamels absolutely clean, a few drops of nitric acid are added to a pint of water and the enamels are then re-washed in this solution. A final washing with clean water is required to rinse away the acid water. I have found that unused wet enamel powder tends to break down and become discolored if stored.

The first coat of enamel is hard fusing to prevent that coat from bubbling through subsequent coats of medium and soft fusing enamel. I wet pack damp enamels with thin to medium size horsehair brushes. I also use a curved dental tool, the double-ended stainless steel kind that has one end shaped like a little scoop. The frame, of course, is not enameled. I wet pack the back first, add a few drops of uncut Klyr-Fyre on the enamel and then draw off any of the excess water with a piece of tissue. When the enamel has partially dried or become tacky, the piece is turned over to enamel the front side. There I wet pack Thompson's silver flux #757, especially if I plan to use transparent reds or pinks; otherwise, I sometimes wet pack the transparents as the base coat. Though leaded enamels yield better colors in the reds, I feel safer staying with the unleaded enamels. The piece is placed on a trivet to dry and then fired just to maturity in my 110V electric kiln at about 1350°F. I use a pyrometer, but my sixth sense tells me when to peep in the kiln.

Next comes placing of the cloisonné wires with uncut Klyr-Fyre. For gold wire, I hammer 22k, 16 ga or 18 ga round wire into the rectangular thinness I want. The wet packing, drying and firing is repeated three to five times until the fired enamel is slightly higher than the frame rim.

Finishing is done by stoning under water, first with a coarse carborundum stick and then a medium-grain one. The grinding continues until all enamel is removed from the frame and the wires, leaving the enamel even. The dull enamel surface is thoroughly washed and glass-brushed to remove any grains of carborundum. For a final finish there is a choice. The surface of the enamel may be rubbed with fine paste wax polish on the fingertips for a matte finish, or the piece may be flash fired in the kiln for a glossy surface.

If the piece is flash fired, the exposed sterling silver frame will need to have the oxidation removed. This removal is a delicate matter because the enamel should not be scratched. I use tripoli or white diamond compound on a hard felt buff on a regular polishing motor. I strongly suggest that if you have not used a polishing motor that you either take a jewelry class or have someone knowledgeable show you how to use the equipment. You can be hurt badly if you are inexperienced or careless. Care must be taken not to overheat the piece with the polishing wheel or else the enamel will crack. If you remove the piece from the wheel when your fingers become hot, you will be safe. When the oxide has been buffed off, a soft flannel buff charged with rouge is used for a final polish of the sterling silver frame. If the piece is a pin, the pin stem is attached to the findings that were soldered on before enameling. For small surfaces and getting into crevices with buffing wheels, I use the Dremel electric tool following the same progression of buffing wheels with white diamond, tripoli and red rouge.

For plique-à-jour earrings, I bend and hard solder sterling silver 14 ga round wire in the shape of a fish or a bird and then flatten the forms with hammers. Sometimes I solder on a jump ring before the depletion gilding. Sometimes I drill a hole for the ear wire. The sterling is prepared for enameling as previously described. After hot pickling and rinsing the frame, it is placed on a sheet of mica and then on the firing planche. The enamel will not adhere to the mica. The enamel is wet packed with transparent enamels, dried and fired. This process is repeated until the fired enamel is at the edge of the rim of the frame. The piece is stoned gently under running water until no enamel is on the frame. The piece is flash fired, and the sterling silver frame is polished after the mica sheet is removed.

I have always had a small fire extinguisher in my studio, but, fortunately, I have never had to use it. One needs to respect and approach with caution acids, electrical equipment and other tools to be able to work safely. ■

FOILS: FINE SILVER AND 24K GOLD

MARIANNE HUNTER

Marianne Hunter had been a painter and crafts explorer when she began enameling in 1967. She is self-taught. After 12 years of discipline in the grisaille technique, she turned to color and foils that gave her a feeling of limitlessness. Her metalsmithing techniques are her own, the necessity of concept driving the ability to execute. Everything about her work is personal; it is driven by passionate yearning. Enamel work has been her only employment since 1967.

My jewelry incorporates elements of enamel on fine silver or copper. The completed pieces range in size from about 2" x 3" to 16". The copper is 18 ga to 20 ga and the fine silver is 20 ga to 24 ga. I begin each piece with an exact drawing to scale. I need to plan because most of my work is a combination of materials and they all have to fit together. The drawing also allows me to plan for placements of gold and silver foils over and under layers of transparent and opalescent enamels. I only make color drawings for clients; otherwise I see the color in my mind's eye. Each piece is a story I tell myself as I work.

I still use the Trinket Kiln, essentially a hot plate, I purchased in 1967. Each enamel piece takes from one to even 30 firings. I first enamel the front of the piece. Generally, I start with Thompson's #124, 80 mesh, leaded, medium black enamel sifted through a piece of silk stocking held in place with a rubber band or binding wire on a small vial. The first layer is a full coverage, slightly thicker at the edges to protect against burning out. This firing is to orange peel to be able to check for flaws and open pits with a file. The subsequent layers are fired just to maturity, being careful not to over fire. I re-sift and re-fire up to four times until I have a base coat that feels just right.

The images are built up in very thin layers of enamel, applied dry by sifting or by laying on with a tiny knife. I use 24k and fine silver foils cut into precise tiny shapes according to the drawings. The foils are placed over parts of previously fired layers and then covered with successive layers of transparents. The enamel piece requires from 12 to 30 firings because of the thin applications of transparents over the foil.

Since I am contact firing the piece on the surface of the Trinket Kiln, I put off counter enameling as long as possible until I feel there is a danger of cracking. To counter enamel, I sift a layer of leftover enamels over a brushed on light coat of AAMCO enameling oil. Originally, I used mineral oil and that worked fine. I do not remember why I changed. I use a flat sable brush to apply a light coat of oil under the enamels and the foils. I tap off the excess enamel and fire the piece, right side up on a trivet. If that is the last firing, fine. If not, each successive firing must be on a trivet with appropriate adjustments in timing. If I have forgotten, in the excitement of the piece, to use the trivet and I have to pry the piece loose from the hot plate, I do it as quickly as possible.

The foil is not added until the design cannot proceed without it. After the front base coat is perfect, I may build up 200 mesh white enamel layers in grisaille, which may then be followed by some areas of foil. I use both 24k gold and fine silver thin foil extensively and cut the foil with a #11 Exacto knife blade. I place the foil on either a piece of smooth cardboard or old postcards with tracing paper on top of the foil. It is important to hold the materials taut and to cut away from where you are holding it. For the large areas of foil, 1/4" x 1/4" or more, I pierce the foil a few times with a sharpened needle; I have forced the eye of the needle into the eraser of a pencil. Too many holes will increase the possibility of tearing the foil. The foils are positioned with oil, dried, fired just to fusing, cooled, and then covered with silver flux. The last firing is the

"Kabuki Kachina Mondrian Midnight," copper base, 24k and .999 silver foils, enamel is 2.5" x 2.5", opals, princess cut diamond, abalone pearl, tanzanite beads, 22k gold, and apatite. Setting: 24k and 14k gold, sterling silver fabricated, embossed and engraved. *Photo by G. Post*

"Firebird" pendant/brooch, grisaille and foil. 3.25" x 2.125" x .06", enamel on 18 ga copper, 24k gold wire, opals, emerald, ruby, frame: fabricated 14k gold, sterling silver. *Photo by G. Post*

"Kabuki Kachina Listens" pendant/brooch, grisaille and foil, enamel on 18 ga copper, 24k gold wire, agate, carved carnelian, frame:14k gold fabricated. *Photo by G. Post*

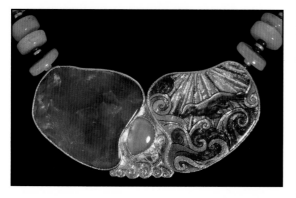

"The Light Caresses" pendant/pin, 3.5"x 1.5"x .25". Foils, enamel on 18 ga copper, 24k gold wire, frame: fabricated, 14k gold, sterling silver. *Photo by G. Post*

"Rhoma's Garden" pendant/brooch 3.5"x 2.36"x .06", foils, enamel on 18 ga copper, 24k gold wire, peridot, emerald, tourmaline, sphene, frame: fabricated 14k yellow and green gold. *Photo by G. Post*

final step for the enamels. Since the layers I use are so thin, any stoning etc. would only remove detail and serve no purpose. The "modeling" of the surface, achieved by the layered building up of the images, is an integral part of the overall feel of most of my work.

The metal jewelry work is individually hand-fabricated and engraved. I use 24k for all the bezels and about 90 percent of the other soldered decorative appliqués. I use 14k gold and/or sterling silver for the supporting structure and clasps. Most of my bezels are textured with an engraver to reflect light and emphasize the gold color while adding to the overall richness of the piece. When all the sections are finished and assembled, I engrave on the back of the assembled piece my signature, the number of the piece, the date, and the title with a poem.

My best advise? Don't follow rules (except safety ones). Don't follow me or anybody else! Use information for inspiration for your own experimentation. If your work reflects someone other than yourself, you will have fun as a hobbyist, but as an artist you will be wasting your time. Bring your own voice to the song of visual language.■

"Kabuki Kachina of the Dawn Rose,"pendant/pin, 1.84"x 3.18"x .3", grisaille and foils, enamel(under-fired texture) on 18 ga copper, 24k gold wire, natural gold crystal, Montana sapphires, cat's eye tourmaline (rare), diamond in white gold, rose petal pearls, frame: fabricated. *Photo by G. Post*

Champlevé and Basse-taille on Fabricated 18K Gold

RICHARD MCMULLEN

Richard McMullen

Richard McMullen has been dealing in fine antique jewelry for more than 30 years. His love of the Art Nouveau period lead him to study the metalsmithing and enameling techniques of the master jewelers of the early 1900s. He has been perfecting these same skills over the past ten years to create his own unique designs.

Virtually all of my inspiration comes from the art nouveau period, when the natural form was elevated to center stage by the great masters. Enamelists such as Lalique, Vever, and Fouquet primarily used the basse-taille technique. This process lays transparent enamel over a chased or engraved metal. I usually make pins, pendants and earrings within the 2.5" size. My jewelry is fabricated of 18k yellow gold with the champlevé cells made by a burr in the flexible shaft. Most of my pieces are one of a kind.

Having drawn and then traced my design on quality tracing paper, I paste the tracing to the 18k yellow gold sheet that I have fabricated and rolled out. The enamel areas are gutted out with a cone burr in the flexible shaft. Those depressed areas are chased; the piece is cleaned and enameled. The completed fired enamel fills about 75% of the cell depth.

I start by fabricating an 18k yellow gold sheet using 75% of .999 yellow gold casting grains, 12.5% of .999 fine silver and 12.5% of .999 copper. I usually start with 40 dwt of gold grains. For melting, I use a gas/air Harris torch with a large tip and a reducing flame that covers the whole crucible. For soldering, I use a mini-torch with a #0 or #1 tip. The metals are melted in about a 3" diameter ceramic crucible; the pure gold is first, and then the silver and copper are added. The torch is kept on the crucible at all times to keep out the oxygen. When all is melted together, I stir the molten metal with a graphite rod and pour it into a vertical ingot mold. The ingot is rolled down to 1.4 mm.

I trace my design onto a good quality tracing paper and then use a water-based glue to adhere it onto the gold sheet. When the glue is completely dry, I use an Exacto knife to trace the internal lines that will serve as the walls for the enamel. This leaves fine lines on the gold metal sheet that I can use as a guide for the chasing design. Next, the outside shape of the piece is cut with a #2 jeweler's saw blade. The tracing paper is removed with hot water, and the edge of the piece trued with files. Then, with an extra fine point Sharpie permanent marker, I draw over the internal marks that I made with the Exacto knife.

Now the fun starts! With a 1 mm inverted cone burr in the flex shaft, I begin cutting a deep groove following the lines I made with the marker. Next, I use ball burrs of various sizes to gut out the cavities to a depth of about 0.3 to 0.4 mm. This leaves a minimum of 1mm of gold as a base for enamel. Since I do not counter enamel, I keep the thickness of the enamel to about 0.3 to 0.4 mm to avoid cracking. Many sizes and shapes of burrs are on the market, and with practice you will find your own method to countersink the metal. I leave the cavities with a flat bottom and a consistent depth.

After gutting out the gold and refining the cell walls, I chase a texture onto the bottom of the cells with the piece embedded in a pitch bowl. The pitch is melted in the bowl with the large torch using only gas. Then the gold piece is placed on the hot pitch, heated with the torch until the pitch climbs up the sides of the gold. If the piece gets too hot, it will sink into the pitch. When the piece is cool to the touch, I use my chasing tool, which resembles a small chisel, and a chasing hammer to tap my desired texture into the gold. To remove the piece from the pitch, I heat the piece with the torch (gas only) and lift it out with tweezers. Then, I either soak it in

"Tribal Elegance" necklace, 6.5" x 5.25" x .055", champlevé enamel on handcrafted 18k gold, shaded basse-taille, 51 pavè set diamonds, marquis orange sapphire, and 12" handmade chain.

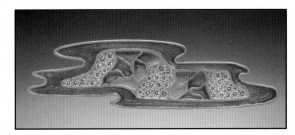

"Reflections of a Sunset" brooch, 2.25" x 1.25" x .055", champlevé enamel on handcrafted 18k yellow gold, shaded basse-taille, plique-à-jour, 41 pavè set diamonds.

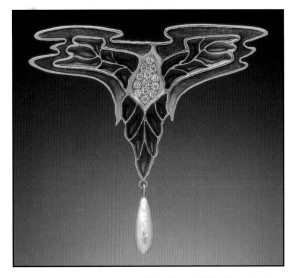

"Aqua Morph" brooch, 2.75" x 2" x .055", champlevé enamel on 18k handcrafted yellow gold, shaded basse-taille, plique-à-jour, 13 pavè set diamonds, fresh water pearl.

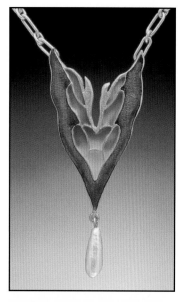

"Nature's Window" necklace, 2.5" x 1.5" x .055", champlevé enamel on handcrafted 18k yellow gold, shaded basse-taille and plique-à-jour. 18" handmade chain.

"Nature's Counterpoint" brooch, 3" x 1" x .055", champlevé enamel on handcrafted 18k yellow gold, shaded basse-taille, 21 pavé set diamonds, pink sapphire.

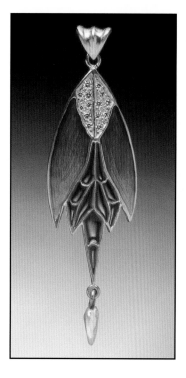

"Autumn Morph" pendant, 2.5" x .75" x .055", champlevé enamel on handcrafted 18k yellow gold, shaded basse-taille, plique-à-jour, 14 pavé set diamonds, fresh water pearl.

turpentine overnight to remove the pitch, or use a torch with oxygen to carefully burn off the pitch. The burning creates a lot of smoke, but with proper ventilation it is the quickest way. I dome some of my pieces very slightly with a plastic-jawed pliers, so as not to mar the surface.

Once the piece is clean of pitch and soaked in hot Sparex2 to remove any oxidation, I solder any necessary pin findings or other accessories with 18k yellow gold medium solder. I once again soak the piece in Sparex2 to remove any oxidation and rinse with clean water. Before enameling, I brush the entire piece under running water with a clean glass brush and Ivory liquid soap. If water beads on the piece, I keep brushing. If repeated glass brushings do not work, I use saliva, nature's perfect cleaner.

When the water sheets, the piece is clean, and then I prepare the enamels. I purchase the standard 80 mesh transparent Japanese leaded enamels. About a teaspoonful of enamel is ground to a fine powder with the pestle in a mortar with enough distilled water to barely cover. Each enamel is placed in a baby food jar and washed with distilled water by filling the bottle half way, pouring off the milky substance and repeating until the water is clear. Usually six or seven washings are sufficient. The consistency of the enamel is now a thick slurry. I scoop about the amount of each enamel I need into plastic spoons and cover each spoon with a plastic cup slotted for the handle to keep the enamel clean.

I fill the cells with a 3/000 sable brush, 3 grains of enamel high. I do not apply a base coat of flux under the transparent colors. After completing each cell, I vibrate the piece slightly by scraping a ribbed dental tool along the side of the piece. This action settles the enamel in place. I leave the outside borders as a frame. Then I wick away excess water with a piece of bathroom tissue or blotter paper. The piece is then placed on a firing trivet either on top of the kiln or under a heat lamp. The kiln is set to 1425°F and preheated for 90 minutes. This preheating prevents large fluctuations in temperature when the kiln door opens and closes.

My kiln is a Vacela with a Thompson digital pyrometer. The safety against over-firing makes the pyrometer worth purchasing. Inside measurements are 11"d x 9"w x 6.5"h. The first firing time determines the total firing time needed for a particular piece. A large analog clock with an easy to read second hand hangs on the wall above the kiln. I note the time of the second hand as soon as I close the kiln door with the piece set inside. I keep the fork in my hand while the piece is in the kiln. I peek in after a minute. If fusing has begun, the enamel has started to darken in color, I give it 15 seconds more. If it is not fused as I want, I put it back in the kiln immediately and give it 30 seconds more. When fired, I let it cool on the top of the kiln for about five to ten minutes so it is cool enough to touch, and I then repack and re-fire additional coats.

When satisfied with my enameling, I clean the borders and cell walls of stray enamel by using a pointed aluminum oxide burr in the flex shaft while frequently dipping the piece in a bowl of water to keep the enamel wet. Any wavy areas are smoothed out with fine diamond paper. All grinding and refining of the surfaces of the enamel are done with water. The surface of the gold is smoothed with a bullet shaped, rubberized abrasive from Swifty Company. This smoothing eliminates the scratches left in the gold by the aluminum oxide burr.

As a last step, I examine the piece with a 10x loupe to ensure its perfection. If it is satisfactory, I glass brush the piece under running water, rinse it again with distilled water and set it aside to dry. Then comes the final firing that is a little longer to totally mature the enamel. When the piece is cool, I soak it overnight in a kosher salt and white vinegar, solution, which is slow acting, to remove the oxides from the gold. The solution is made by half filling a pint jar with white vinegar and then adding tablespoons of the kosher salt until saturation point is reached. Any excess salt remains on the bottom of the jar. I like the matte finished look, so I never use a polishing wheel. At most, I polish it up with a glass brush.

This method is just one of many ways to enamel. With patience, perseverance, and a lot of common sense, the art of enameling is very rewarding for me and can be for you, too. ∎

FINE SILVER 36 GUAGE FOIL

AVERILL SHEPPS

A Smith College graduate, Averill learned enameling in a workshop taught by James Frape. Self-taught since then, she has been creating enamels for 45 years and supporting herself by selling her work at major craft shows and galleries as well as by some teaching. She has developed her own innovative methods and works with transparent colors on vessels, wall pieces and jewelry.

Averill Shepps in her studio.
Photo by Anton Shepps

I have been working in enamels for over 40 years and during most of that time have made a living by selling my enamels. My work consists of one-of-a-kind pieces for exhibition in addition to other pieces and multiples mainly for sale. Included are panels, vessels, both large and small, and jewelry. Most of my enamels are on copper with some on fine silver. I primarily use the sifting technique because of the large size of many of my pieces and because the enamel can be applied quickly. Producing multiple enamel pieces in quantity at one time demands efficient working techniques.

While the larger pieces are usually on 18 ga copper (.04"), I also work on foil sheet in silver and copper as thin as 36 ga (.005"). I use some lead free enamels but most of my pieces are done with leaded enamels. I purchase 80 mesh enamels and grade sift them to various sizes; the grade I use depends on the technique I use or the effect I desire. As a general rule, my bowls and panels that are enameled by sifting are done with enamels 80 - 325 mesh. Enamel over silver is done with 150 - 325 mesh.

The 36 ga fine silver foil that I use can be purchased from Handy and Harman in a 100 oz order or from Thompson enamel in 6"sq. or 12"sq. pieces. This foil can be used by itself to produce jewelry or Christmas ornaments. Small fluted bowls can even be made with it. I discovered that Handy & Harman manufactured a 36 ga foil that had the appropriate qualities and have been using it since 1991. I sought out and began working with the heavier fine silver foil because I wanted a foil that had less surface texture: the crinkled surface of the thinner foil is not always appropriate for certain designs. The photographs show various examples of my work with this heavier foil.

Shapes of foil are cut with small, sharp scissors and then put through the rolling mill. The tree-like shapes, for instance, are made by first cutting a long and very narrow triangle. The metal curls up when cut with scissors; therefore, it must next be flattened with your fingers by placing the piece on a flat surface and applying gentle pressure.

Slits are then cut down the length of the triangle at third or quarter intervals so that six to eight slits are on one side. The metal will again curl as it is cut. The individual foil pieces defined by the cuts can be pulled and shaped with the fingers until a suitable overall shape is reached. The shaped piece is then put through the rolling mill using enough pressure to flatten and slightly stretch it but not so much pressure to distort it. If a tree-like shape is desired, the side with the cuts in it is put through the mill first. If the other end is held firmly with jewelry pliers while the piece is going through the mill, a straight trunk will be formed. The length and thickness of the trunk is determined by the length and width of the original foil shape.

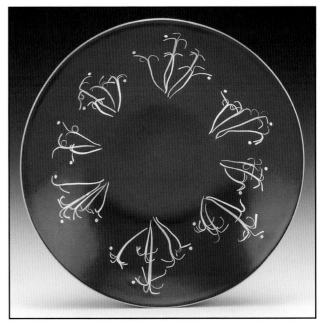

Plate, 7.25" diameter x .875", sifted enamels on copper, rolled .005 fine silver. *Photo by Jerry Anthony*

"Trees With Purple" framed enamel, 6"x 10", sifted enamels on copper, rolled .005 fine silver. *Photo by Jerry Anthony*

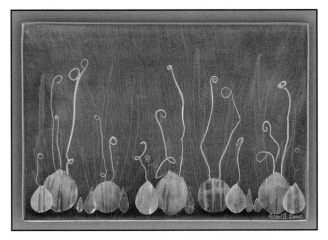

"Bulbs in Lilac and Pink" framed enamel, 5" x 7", sifted enamels on copper, rolled .005 fine silver. *Photo by Jerry Anthony*

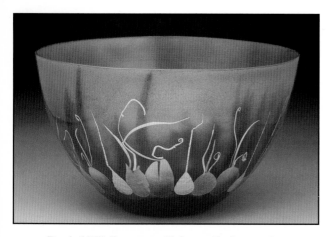

Bowl, 6.75" diameter x 4" deep, sifted enamels on copper, rolled .005 fine silver. *Photo by Jerry Anthony*

The foil shapes are now ready to be fired onto the metal form that has fired enamel on it. On a flat metal surface, I use Klyr-Fyre 2:1 with distilled water; on a sloping surface, I use lily root powder that is mixed to a paste with a drop or two of distilled water. After the adhesive has dried, the piece is fired in a kiln that is no hotter than 1450°F and for only one minute or just until the enamel has softened enough that the silver is bonded to it.

When the piece is cool, enamel is sifted over the silver shapes. Clear flux for silver is especially appropriate because it has the advantage of not changing the color of the base metal. Grade sifted enamel to 150 - 325 mesh will give maximum clarity. The piece is again fired so that the enamel melts and fuses to the silver foil. Care must be taken to be sure that the temperature is no greater than 1450°F for this firing as well as any subsequent firing. The fine silver is very conductive of heat and if it gets too hot, it will go into solution and become part of the enamel. The result will be a cloudy or milky surface. Although the piece may be saved by covering the cloudy area with opaque enamel, subsequent firings will have to be done at a lower temperature to prevent more of the silver going into solution.

This technique requires skill and care, but the results are unique. Putting the silver shapes through the rolling mill gives them added grace and form. The silver shapes are more distinctive because the foil is flatter and does not have the crinkled texture that happens with the thinner foils. ∎

Three Brooches, 1"x 1.5"each, sifted enamels on copper, rolled .005 fine silver, 24k gold granules. *Photo by Jerry Anthony*

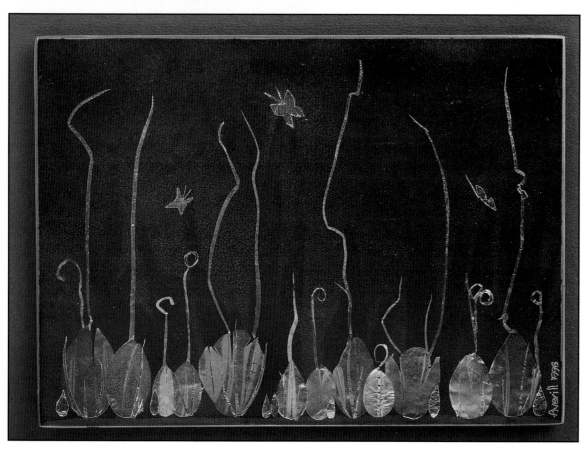

"Spring Bulbs and Butterflies" framed enamel, 5" x 7", sifted enamels on copper, rolled .005 fine silver. *Photo by Bob Barrett*

ENAMELING ON FINE SILVER METAL CLAY

JEAN VORMELKER

Synthesizing an eclectic educational background, Jean Vormelker brings a unique perspective to her jewelry and enameling work. She has studied in the Untied States, Japan, Australia and Canada, and is responsible for her own development. A multi-faceted person, skilled in metal and enameling techniques, Jean's latest passion is working with and teaching workshops in PMC (Precious Metal Clay) and enameling.

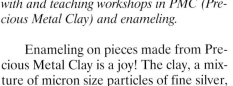

Jean Vormelker laying out a display at Red Deer College.

Enameling on pieces made from Precious Metal Clay is a joy! The clay, a mixture of micron size particles of fine silver, an organic binder and water, is malleable and will take any shape or texture. It can even be mixed with enamel to form a glass/metal alloy. Handling is similar to working with porcelain clay. There is no waste; all unfired clay scrap is reusable one way or another. Unwanted fired pieces can be melted in regular casting or otherwise recycled.

As with ceramic clay, you make a piece, dry it, and fire it. It is much easier and quicker to finish the piece completely in the leather hard stage by lightly sanding all rough edges. Once the piece is fired, the metal requires heavier grinding and filing to smooth it. No spruing is required as it is with lost wax casting, so if the piece is well finished in the clay state, only some burnishing is required to complete it after sintering.

The sintering process burns off the binder and water and compresses the minute metal particles into solid metal. Sintering is accomplished by placing the dry clay in a cold kiln, bringing the temperature up to a specific setting, 1110 - 1650°F for a specified time. The temperature used depends on which formulation of the clay has been chosen and what inclusions are involved. Shrinkage occurs in all directions during sintering, but what you see before firing is what you get afterwards, only smaller. The metal will assay as solid .999 pure silver (or gold) and is usable like fabricated metal in a variety of enameling techniques. The ultimate joy for enamelists is that metal clay has no copper to oxidize and cause firescale problems. With the possibility for unusual shapes, pieces can be prepared for champlevé, basse-taille, cloisonné, or plique-à-jour without etching or soldering. The fired piece can be gently reshaped by hammering if it is needed.

I have worked with various formulations of both Precious Metal Clay (PMC), made by Mitsubishi Materials Corp. and Art Clay, made by Aida Chemical Corp. Both brands are available in different basic forms: lump form, currently in three formulations with different firing schedules and density, paste, slip/syringe and paper/sheet. With the consistency of old fashioned window putty, the clay forms work like ceramic clay and are placed between plastic sheets to prevent excessive drying when rolled into sheets. The paste and slip formulations are useful in attaching multi-part pieces and for decoration. The PMC+ Paper is very flexible and can be folded and unfolded like origami. Unlike the other forms, it only sticks to itself when dampened slightly with water or with a tiny amount of white glue. It can be cut into strips for cloisonné and plique-à-jour, draped or used as a base for more intricate work.

The Midas touch can be achieved by using Gold PMC (24k gold lump form) or 22k Gold Art Clay. I mix my own 18k from 24k and silver PMC.

"Eve's Apple" pendant, 1.5" x 1.5" x .315", sifted/wet-packed enamel on fine silver PMC from an original mold. *Photo by Jim Vormelker*

"Artifact" necklace, 3.25" x 2.75" x .25", large element 2.625" x 1.125", enamel on molded fine silver PMC, wire wrapped fine silver frame, recycled vintage simulated green pearl, and wire. *Photo by Jim Vormelker*

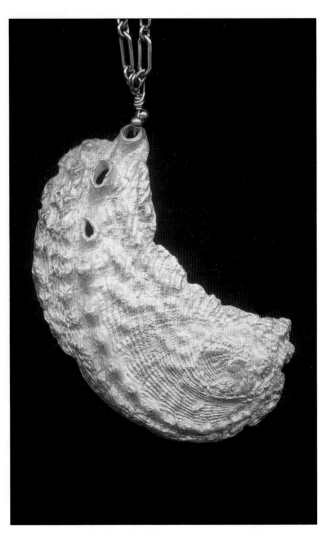

"Golden Eye" pendant, 3.25" x 2.75" x .25", wet-packed surface highlights, unleaded enamel on molded fine silver PMC, 24k yellow gold PMC, 18k green gold PMC in the "Eye." Chain: 14k GF and beads. *Photo by Jim Vormelker*

The lump form can be used for solid gold pieces or thinned to make a wash to be applied in a painterly manner to either raw or fired pieces and re-fired. Traditional Kum-Boo can also be applied to fired silver pieces. Introduced in 2004, PMC Aura 22 (22k slip form) can be used to add a wash of gold to a piece of fired silver and re-fired. Metal clay is a growing field with exciting new formulations added periodically.

Enamel can mimic gold as well. My piece, **"Golden Eye,"** is a combination of 18k green gold and 24k yellow gold PMC worms placed side-by-side, shaped in a spiral and rolled to make a sheet. The sheet was pressed in a mold of an abalone shell. Gold sinters at 1830°F for two hours and so that part was made first. Then silver PMC was molded and a hole cut to insert the fired gold piece. With the next firing at 1650°F, the shrinkage permanently trapped the gold without soldering. Then I wanted to bring the gold colors up into the silver to highlight the undulating surface. After a little experimenting, I blended two "almost right" colors of transparent enamel for each color of gold to give me the colors I needed. I mixed them intuitively and got it right. These enamels were wet packed on the raised portions of the shell in a painterly manner and fired normally.

Metal clay sold in the United States is either fine silver (.999), fine gold (24k PMC) or 22K Art Clay Gold. The gram designation on the package is the amount of silver (or gold) in the package. One ounce of original PMC is about the size of the end joint of a woman's thumb. PMC+, PMC3, and Art Clay have more silver, less binder and water per ounce of clay. Therefore, an ounce of these is smaller in size and comes in different size packages, too. I find the best price by breaking them all down to a per-gram figure, rather than going by the package price, because of the different packagings available.

How to actually produce a silver or gold object with metal clay is adequately covered in other literature. For earrings, pins, pendants, necklaces and other light stress or larger objects, my choice is the original PMC formulation. The pieces are relatively light weight compared to fabricated forms of the same metals. The shrinkage of 25-30 percent allows detail that could not be done any other way. It is the most porous formulation and fires at 1650°F for two hours.

Better choices for rings and bracelets are PMC+ or Art Clay Silver with 10-12percent shrinkage. They are more dense and will better withstand abuse. PMC3 (also 10-12 percent shrinkage) has the finest grain structure and is best for heavy wear when fired for two hours at 1650°F, even though it can be fired as low as 1110°F for 30 minutes. When working to take advantage of the different shrinkage rates, a sheet of original PMC as a base with woven PMC+ Paper strips on top creates a natural doming of the combined flat sheet when fired. (Basket Weave)

Sometimes I sketch an idea, but I rarely draw a design in detail. Part of my inspiration comes from working intuitively and directly with the material. I make press molds of my found objects from RTV (Room Temperature Vulcanizing) material, roll patterns in the clay and manipulate the clay in a number of ways. Hollow forms are easily made with a core of cork clay that burns away during sintering. In the leather hard or dry stage, the rough spots are sanded smooth with salon boards or finishing paper and then fired.

I use a Vcella 240V enameling kiln, 12" x 12"x 6.5", and a SC-2 Paragon 8 "x 8"x 6" controlled kiln for PMC on a 120V line. The controlled kiln can also be used for enameling if all traces of alumina hydrate are vacuumed out. Alumina hydrate, used to support contoured PMC during sintering, does nasty things if it gets on the enamel. To further avoid cross contamination, I also use different shelves for PMC and enameling. The PMC kiln has space age insulation and looses heat more rapidly when opened than the Vcella fire brick does, so I run it a little hotter than normal for a faster recovery time.

After sintering and before enameling, consideration needs to be given to the differences among the ways the light reflects into transparent enamel from a natural white matte surface fresh from the kiln, from a brass brushed satin finish and from a burnished highly polished one. A variety of finishes adds visual depth to the piece. The same color will look darker over a matte finish and lighter over a shiny one, but to have a consistent color regardless of how the light hits it, the surface must be finished evenly whether it is matte, satin or burnished. This finishing is particularly important on smooth surfaces where every flaw is noticeable. I like the look of just burnishing the raised surfaces and, when not enameling, I sometimes oxidize the low areas.

It is hard to go back to the original softer finish if the whole piece is burnished to a hard bright shine in the beginning. So I bring up the shine gradually, stopping to evaluate at each step. Taking care not to disturb the areas that are to remain matte white, I usually start with a brass brushed satin finish to bring up the metallic color on the parts to be shined. Then further burnishing closes the pores of the metal and puts on a high shine.

If the surface is not closed by burnishing, excess air bubbles might form in the enamel because of the natural porosity in the metal. I find that using either a rotating or vibrating tumbler with mixed steel shot and needles, a little Fels Naptha soap or non-sudsing detergent and water does an excellent job of burnishing to a consistent overall high shine. The tumbler can run for hours if necessary and doesn't tire before the job is done. I hand burnish to reach recesses that the tumbler missed. My PMC test pieces show that using the tumbler for an all-over burnished surface produces fewer small air bubbles in the enamel than hand burnishing.

Enameling requires a clean surface, so I do not use a buffing machine, which might leave polishing compounds in crevices and then could be difficult to remove. I have used 80 mesh, unwashed, leaded and unleaded enamels in thin coats without discernable loss of brilliance. (Eve's Apple) Counter enamel has not been necessary on thicker pieces when thin coats of enamel have been used. Usual enameling practices apply when thicker enamel is used on thin pieces. I apply the enamel by wet packing or sifting because I like the blending of colors. An adhesive is used only when the shape requires it. The firing is routine on a trivet.

Color samples are invaluable. Enamel colors on silver appear very different from the colors on copper, so I made separate color charts of leaded and unleaded transparent enamels on burnished PMC using both sides of the piece to complete the palette. For the leaded enamels, I made a piece of PMC, finished size 1.5" x 3" x 22 ga, with a tumble burnished finish. Alternating ¼" stripes of bare silver and a base coat of Ninomiya N-1 Flux for Silver were fired normally. Then unwashed dry enamel colors were applied in .25" x .5" blocks covering a strip of flux and bare silver and fired. This color chart tells me which colors need the flux under them. Some colors do not look good either way; others look fine with or without the flux. I love Ninomiya N-1 Flux for Silver. Other fluxes for silver I have used turned yellow, but N-1 stays crystal clear.

The layout for my unleaded color chart is on two smaller pieces of PMC, with one side showing the colors over flux and the other side showing them on bare silver. By placing them side-by-side, I can view the whole spectrum with the same base or flip one to the other side to see the same colors over both silver and flux.

Applying enamel to sintered and polished PMC or Art Clay pieces is the same as enameling on any other fabricated or cast metal, sifting or wet packing 80 mesh enamel as usual. I find that firing temperature should be kept low because 1335°F-1400°F. Higher temperatures may change the color and may put an unpredictable metallic sheen on some enamels or turn them yellowish due to chemical reaction with the silver.

"Basket Weave" pendant, 2.75" x 1.75", modified basse-taille, leaded enamel on fine silver PMC+ paper. Chain: sterling silver. *Photo by Jim Vormelker*

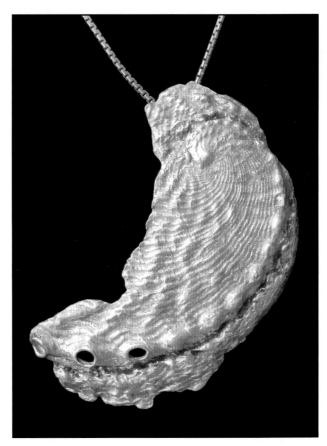

"Blue Abalone" pendant, 2.75" x 1.75", modified champlevé, leaded enamel on fine silver PMC. Chain: sterling silver. *Photo by Jim Vormelker*

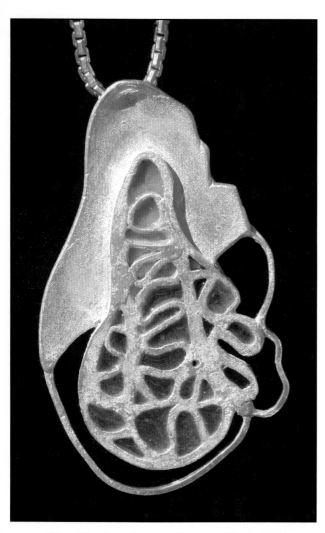

"Blue Maze" pendant, 1.125" x 1.0625", plique-à-jour and cloisonné, leaded enamel on PMC fine silver, fine silver wire. Chain: sterling silver. *Photo by Jim Vormelker*

Finished pieces can stand alone but also can be appliquéd to other enameled forms as in Oceanic. The shells are enameled PMC pieces, enamel attached to a pre-fired engraved and formed copper sheet with enameled round copper wire waves. That is then cold attached to an enameled thin steel sheet with additional PMC pieces added.

I continue to experiment with metal clay. An interesting variation is to apply dry enamel to the unfired dry clay piece before sintering. The piece can be left with the matte surface or the clay can be burnished before applying the enamel. I recommend using either the PMC+, PMC3 or Art Clay for this technique because of their greater density. If these are used and sintered at lower temperatures, the enamel will stay on the surface and leave a clear silver surface on the back. A sample I made using dry enamel on the dry original PMC clay piece before sintering, fired with a lovely matte enamel surface, but because of the greater porosity of original PMC, the enamel sank all the way through and left a stained, blotchy back. It did not show the darker colors usually associated with long, hot sintering; therefore, for this technique, use original PMC only on pieces when the back will be covered or enclosed.

If you plan to fire the enamel longer than the usual two minutes at 1335°F-1400°F, it would be wise to make a color chart using the longer sintering time and lower temperature you plan to use. Look at your first color chart to choose appropriate colors for further testing. Avoid reds and oranges as they darken quickly. For future reference, do keep a record of colors used, firing times, temperatures and results.

Adding color in the clay to make a new metal/glass alloy (sometimes referred to as Rainbow PMC) is another technique with which I have experimented. Before the other formulations were on the market, I played with this technique using original PMC and was not happy with the results. Medium expansion enamels are made to mature in two to five minutes at 1500°F. I had used an equal volume of Thompson's leaded #715-Blue Jay enamel and clay. The same Blue Jay enamel when mixed with PMC+ and sintered for a half hour at 1470°F retained a light blue color as the silver turned to a matte white color. As the temperature was increased to 1560°F-1650°F for 20 minutes and 10 minutes respectively, the color turned a darker olive green. The longer two hour sintering of original PMC at 1650°F essentially burned the enamel almost black. With the greater shrinkage and porosity of original PMC, the enamel sank to the back and squeezed out in lumps. Not a pretty sight and only color tests could have shown me what color change to expect.

PMC+ was better for this technique than original PMC because of lower shrinkage. Now that PMC3 is available, firing time can be even shorter and temperatures much lower so the enamel does not "burn." Nifty mini-measuring spoons for a Drop, Smidge, Pinch, and Hint are available so you can make small quantities for testing and repeatable results. The largest Hint spoon equals 1/8 tsp. Each smaller spoon is half the one before it, down to a Drop. Mix ratios range from 1 part PMC3 to 1 - 4 parts enamel: 1 Drop of PMC3 to 1 Smidge of enamel would give a 1:2 ratio; 1 Drop PMC3 to 1 Pinch enamel would be 1:4. Color density increases with higher ratios. Depending on the ratio used, firing at a low 1300°F for 10 minutes may sacrifice strength to get the color you want. For greater strength, firing at 1450°F for 5 to 10 minutes will make colors more brilliant, but then some will turn muddy at ten minutes, so testing is essential. Firing below 1300°F results in a brittle, easily broken piece.

Choice of colors is important, with blues and greens being the best. Check your original color charts for compatibility. With a mixing spatula, thoroughly mix in only one color of enamel with the PMC3 and add a little water as necessary to maintain workability. Label each color, note the ratio and keep each color separate. Mixing two colors doesn't work well. Form the piece or use the alloy as an appliqué and let it dry. A thin under layer of plain clay on the back will protect the kiln from melted enamel and the piece will be more comfortable to wear.

Sintering for this technique starts in a pre-heated kiln. Place the dry piece on a kiln shelf coated with kiln wash, on mica, or on ceramic fiber sheet to prevent the enamel from sticking to the shelf. Put shelf or trivet in the pre-heated kiln, bring heat back up and start timing at the correct temperature. Depending on how much enamel is used and how well mixed it is with the clay, the piece will have anywhere from a shiny enamel surface to a grainy satin finish with glass throughout the piece. Polish gently. Remember: once fired, the piece cannot be reshaped because of the glass component.

One of the unique qualities of metal clay that I particularly like is its ability to make a seamless blending of different karats of gold. A firescale-free 18K can be mixed with gold and silver PMC. Ropes, rolls or worms of 18k and 24k gold PMC are placed side by side and rolled together to create a sheet of blended colors that is impossible to duplicate any other way (Golden Eye). Although fine metals are too soft for some applications, metal clays with their unique properties are a wonderful addition to the lexicon of metal techniques for the enamelist.■

VESSELS

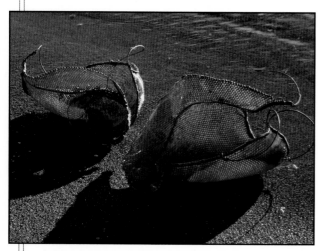

"Time and Tide II" set of vessels, 8" x 14" x 10", plique-à-jour enamel on copper mesh, copper wire, liquid and 80 mesh enamels, glass beads. *Photo by Kate Cameron*

"Sail Me Away I" set of vessels, 7" x 15" x 9", plique-à-jour enamel on copper mesh, copper wire, liquid and 80 mesh enamels, lusters. *Photo by Kate Cameron*

COPPER MESH VESSELS

ALANA CLEARLAKE

Alana Clearlake's first memory is observing the changing patterns of color and light outside her nursery window. Later, her favorite toy was a box of crayons, and the best Christmas gift she ever received was a card table to use for those crayons. She is still fascinated by patterns of color and light, but now the table has multiplied to many larger ones, and the crayons have segued into enamels.

Alana Clearlake

For several years, I have been exploring the potential of enameling on copper mesh, specifically on an Amaco product called *Wireform*. It is an expandable screening material of solid construction, that is, it is not made of woven wire like most mesh products. *Wireform* can be purchased at most art stores. Although it is available in various sizes, I prefer the flat, not folded, 20" x 16" sheets.

I make organic vessel shapes with echoes of Art Nouveau, seashells, flowers, and plants. I also use the mesh as if it were paper by folding it with a modified origami technique and then assembling the folded mesh into vessels. I have loosely followed the technique of unit origami, in which several to many pieces of paper are put together to form symmetrical polyhedrons. As I prefer asymmetry, I have been pushing the technique to make one-of-a-kind vessels instead of an enclosed symmetrical form based on someone else's pattern.

In both cases, I do not start with a sketch on paper, but with a visualization that I work toward. This method is in direct opposition to the way I work on my jewelry and wall pieces, which I do sketch first, often using pencils to define the color pattern. I do not know why I work differently 3-dimensionally, although I was encouraged by college professors to pursue sculpture. Perhaps they sensed that this technique was a more "natural" fit for me.

ORGANIC MESH VESSELS

First I will describe my method of making the organic mesh vessels, which range in size from approximately 3" x 3" x 3" to 12" x 12" x 12". The lightweight quality of *Wireform* makes forming easy, but the weight of the enamel on larger pieces causes the shape to distort when fired. Any piece with a dimension of 7" or greater needs to be electroformed after shaping to strengthen the mesh.

I begin with an idea for a series of similar forms, such as tall and narrow, or mostly closed with a small aperture. Wearing leather gloves, I cut shapes from the mesh using titanium scissors. I use the rounded end of a plastic hammer against a sandbag to form the mesh into a three-dimensional shape. Then I reinforce the edges by rolling the mesh over 12 ga to 20 ga copper wire and I use the extended ends of the wire to tie the form onto a lightweight trivet (described below) for firing.

The pieces are pickled in a Sparex2 solution. I use the directions on the can to mix some into solution in a kitty litter plastic container. I add some more Sparex2 when the solution starts to get weak. I date the container when I refresh the solution. If the volume gets a little low I add more water. The Sparex2 solution I am using is now three years old. It does not seem to matter what the strength is; it just takes longer to clean as it gets weaker.

After pickling, the piece is rinsed, dried, and then dipped into semi-clear liquid enamel (Thompson's water base, liquid form, clear transparent #BC 303L). Holding the vessel by the wire extensions, I sift powdered enamels onto the wet enamel. I use mostly 80 mesh unleaded enamels after sifting out the fines to use for painting enamel. I sometimes use leaded enamels as well.

The trivet I use consists of two lightweight trivets with legs that are wired together. One trivet is turned upside down, and then the two trivets are wired together using 22 ga copper wire where the three triangular steel pieces of each trivet meet. These triangles are normally the supports used for firing. The vessels are hung, usually right side up, by tying the wire extensions onto the metal bars of the upper trivet. After attaching the piece, I let it dry completely before firing. Since *Wireform* is lightweight, firings need to be of short duration so that the mesh does not distort when heated. The first firing is quick: 30 to 45 seconds at 1450°F.

I have an indoor/outdoor studio in California. My kilns, polishing equipment, electroforming tank and torch are outside under a tent. Great in the summer! Everything is stored in plastic containers with lids. I have two kilns that I use: a 110 Vcella with an interior size of 7" x 12" x 12," and a 220 Vcella with an interior size of 15" x 16" x 16". Both kiln floors are protected with furnace floor blanket and have digital controls. The size of the object determines which kiln I use, but the 7" height of the smaller kiln is a real limitation for firing three-dimensional forms.

After a vessel cools, I remove it from the firing rack and check the shape. Some readjusting can be done at this point with the plastic hammer and sandbag. I wear glasses to protect my eyes from flying glass shards. After the second firing, it becomes much harder to change the basic form, and disaster can occur if any change is attempted because the mesh cracks and the fissures are difficult to repair.

After any necessary reshaping is completed, I spray the form with Klyr-Fyre, diluted 50 percent with water, and sift on more enamel. I use a Preval sprayer that I have figured out by trial and error how to keep from clogging: use no thicker than a 50/50 solution, fill just below the fill line, spray in short bursts. If the sprayer stops working: stop spraying right away, remove the cartridge, place the tube into a cup of water and spray to clear the nozzle. The piece is re-tied to the rack and re-fired at 1450°F for a slightly longer time. I continue in this manner for one to four more firings. I have tried to fill in all the holes in the mesh by repeated firings, but too many heatings cause impurities to leak out of the mesh and bubbles to appear in the enamel. The limit for success seems to be six firings, all around 1450°F, increasing the time with each firing for a maximum time of around 1½ minutes. When firing is completed, I straighten the copper wires and integrate them into the design of the vessel. I sometimes "ball up" the ends of these wires with a torch.

This technique requires experimentation, since the finished result depends on many variables: the thickness of the liquid coat, under and over firing, thin or thick application of the enamel. A result is not wrong, just a visually different one. The joy of this technique is progressing from an idea to a finished form rather quickly.

UNIT ORIGAMI

Using the copper mesh as folded paper is not such a fast, direct method. First, it is necessary to learn origami. Although many of us have made an origami bird at some point, unit origami is a more sophisticated method of constructing a geometrical shape that is based on 30°, 60°, 90° right triangles, or 60° equilateral triangles. The paper is folded so that there are tabs and pockets into which to insert the tabs. I discovered that paper

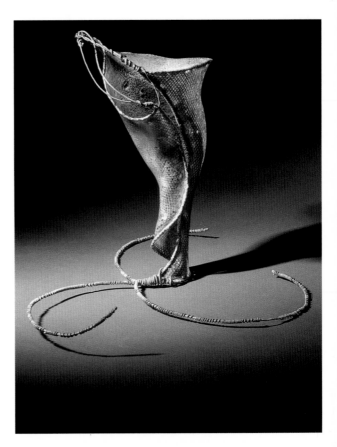

"Calyx I" sculpture, 14"x 8"x 6" plique-à-jour enamel on copper mesh, copper wire, liquid and 80 mesh enamels, lusters, patina. *Photo by Kate Cameron*

"Origami Series: Apertures" vessel, 6.5" x 7" x 5", liquid and 80 mesh enamels on copper mesh, folded, riveted, electroformed. *Photo by Kate Cameron*

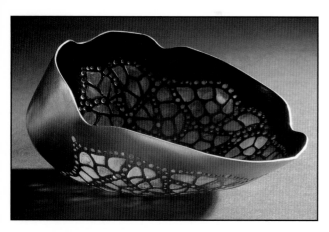

"Peace" vessel, 3 "x 5" x 4", pierced plique-à-jour, wet-packed enamels on copper, patina. *Photo by Kate Cameron*

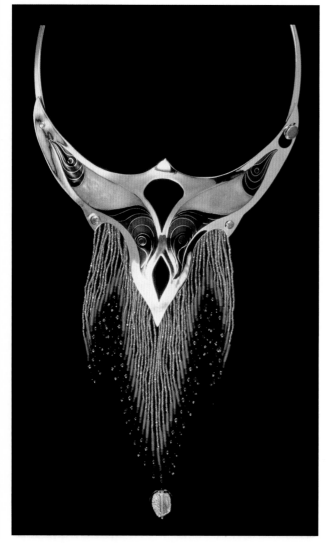

"Wings Learning How to Crawl IV" neckpiece, 12" x 6" x 6", cloisonné, champlevé, enamels, hand formed, silver, seed beads, opals. *Photo by Jim Tanner*

origami is not a permanent medium: I have observed my practice forms begin to self-destruct, surrender to gravity, and collapse.

The *Wireform* substitutes fairly well for paper, but it does stretch and does not re-bend in a reverse direction very successfully. Very much of origami is based on reverse folds so the technique has to be adapted to the mesh as well as to 36 ga copper foil, which I use along with the mesh. Basically, sections of mesh or foil are cut off, and folds are made in one direction only.

Electroforming the finished vessel before enameling is almost always necessary to strengthen the overall form. Pretty small forms can be successfully enameled without electroforming, but anything larger than 4" begins to give way when heated as the paper objects do with the passage of time. Most of these vessels range in size from 6" x 6" x 6" to 12" x 12" x 12" and are fired in the larger kiln.

I cut the mesh and foil with a mat knife, using a stainless steel ruler as a guide, on a resealable, rubber cutting mat. A sharp knife is necessary for clean cuts, so a mat knife with break-off blades is useful. Also, unit origami is based on math and geometry, so using the metric measuring system saves a lot of grief. It is much easier to divide 100 mm into halves and quarters than it is 9 7/8".

After the vessel is assembled, I compress the mesh with long flat nose pliers as well as with the wide end of a ring mandrel. I use a telephone book for a support, especially where the tabs are inserted into the pockets. I sometimes add rivets to stabilize joins that may be stressed when the piece is fired. I use copper or steel pop rivets or ones that I make from 16 ga or 18 ga copper wire. Occasionally, I weave copper wire, 20 ga - 24 ga, into areas that need extra strength.

At this point the vessel is ready for electroforming. The piece is plated for 6 – 24 hours. My electroforming tank is 18" x 18" x 18" and holds approximately 20 gallons of solution. I found the formula for the solution in *Electroplating and Electroforming for Artist and Craftsmen* by Newman and Newman. For more information about this technique, consult either this book or Oppi Untracht, *Metal Techniques for Craftsman*.

After electroforming, I rinse the vessel thoroughly in water with a spray nozzle attached to a garden hose (outside!), then boil it for 5 minutes in a baking soda solution of 1/2 cup in water to cover the piece in a two gallon stainless steel pan, after which I rinse it well again. I dry it instantly and thoroughly with a hair dryer after each rinsing as this eliminates oxidation of the copper. I check for any dark blue color on the form, which is an indication that there is still some electroforming solution trapped inside the joins. The vessel goes into the baking soda bath for a second time and then is well rinsed again. Two of these washings and rinsings seem necessary.

These pieces are also dipped into semi-clear liquid flux (Thompson's BC 303L) and fired at 1450°F for 1½ to 2 minutes. Unlike with the organic vessels, I try to achieve a good coating of the liquid without sifting on enamel in the first and second coats. To fire, I also hang them from a trivet, one that I have had made resembling the one I cobbled together, but that it is taller for larger vessels. For these origami pieces, I add temporary wires looped around the vessel and then tie them onto the crossbars of the trivet.

It is sometimes necessary to pickle the piece between the coatings of liquid enamel in order to eliminate firescale. I rinse the Sparex2 off with water, boil the piece in the baking soda solution, and then rinse well with water again. I also dry them completely before firing. This flux-only method for the early applications allows any trapped electroforming solution or Sparex in the joins to burn out before color and a heavier enamel coat are added.

After a thorough ground coating is achieved, these vessels can be enameled in almost any technique. The first coat can also be any liquid color instead of semi-clear flux. To date, I have not had success using powdered enamel without first using a liquid enamel to penetrate into the joins and seal them. Again, this technique requires experimentation: I have had many failures as I worked toward success. Although I have needed a wastebasket for my efforts, I have also enjoyed very many results! ■

VESSEL FORMS

SARAH PERKINS

Sarah Perkins received her BA in Art from San Diego State University in 1979 and an MFA from Southern Illinois University in 1992. She is Associate Professor of Art and Head of the Metals program at Southwest Missouri State University and a Trustee of The Enamelist Society. Her work has been shown extensively in the United States and abroad and also has been published in Ornament, American Craft and Metalsmith publications.

Sarah Perkins at the workbench

I consider vessels my most important work, but I also make jewelry and utensils. I usually have sketches and color studies of a piece before I start, unless it is one in a series that are all similar and I can envision it easily. I do not do detailed drawings because then there would be little to keep me interested in the piece while making it.

My vessels range in size up to 9" tall, which is the largest dimension my kiln can handle. I have two kilns: one is 8"w x 10"d x 10"h on a 110V line; the other is 12"w x 14"d x 12"h on a 220V line. I prefer kilns that are deeper than they are wide because the heat is more even farther away from the door. A 220V line recovers faster, which is important when firing larger pieces. I never let the kiln go below 1350°F because the metal could oxidize too much before the enamel flows and cause the enamel to adhere to the oxide layer, which does not stick to the metal and will chip off or peel back. Even in later firings, the enamel always cracks when put in the kiln and needs to have the temperature recover quickly so that the enamel can heal properly.

I use primarily fine silver or copper for vessel forms. The larger pieces are 18 ga or 20 ga; smaller vessels are 20 ga to 22 ga. I would make a 5"d x 5"h bowl of 20 ga fine silver. However, if there were to be soldered additions to the bowl, I probably would make it of 22 ga. I do not clean my metal first unless it has blobs of crud on it, which I just brush or wash off. If I am using transparents throughout, I remove any scratches with a burnisher. It is easiest to enamel on seamless copper or fine silver forms, which means that you must either raise or spin the metal form. Of course, you can buy simple spun forms in copper and alter them. If I want to alter the shape of the vessel I do it before I attach a rim and/or a base.

It is also possible to enamel a seamed piece; but as enamel does not adhere well to solder, the seam must get special attention and is a complication. The three ways I deal with a seam are by: soldering a square wire over the seam, the thickness of the wire being the same as the finished depth of the enamel; covering a clean butted seam with standard metal foil over which I enamel; or using a high lead content enamel as a base coat over the seam.

It is helpful to have attached rims and bases to rest the vessels on when firing them in the kiln. The rims also protect the enamel edge and give the piece a finished look. To solder on a protective rim, it is advisable to have it form an overhang on both the inside and the outside of the piece. To ease fitting and soldering, I make the rim wider than necessary and trim it later.

When soldering fine silver to fine silver, I use hard or IT solder with a high temperature white paste flux containing fluorides, such as Handy

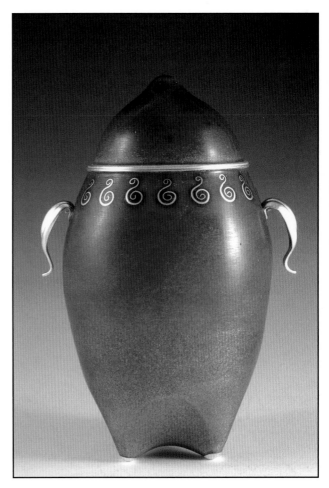

"Burned Beggar's Bowl" raised copper bowl, 3" x 5" x 5", enameled.

"Winged Container" 2004, raised, fabricated container, 6" x 3" x 4", cloisonné enamel on fine silver. *Photo by artist*

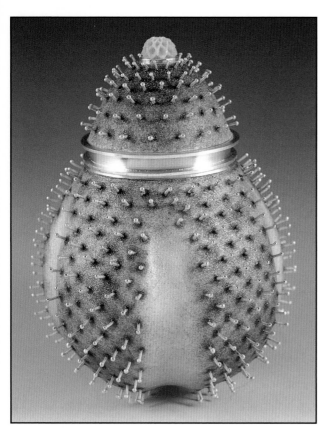

"Cactus Container II" 2004, raised, fabricated container, 5" x 4" x 4", cloisonné enamel on fine silver, coral. *Photo by artist*

"Mae" teapot, 5" x 7.5" x 4", raised, fabricated, enamel on fine silver, ebony, citrine. *Photo by artist*

Flux. This flux gives off toxic fumes and should be used in a well-ventilated area. Unfortunately, the safer fluxes do not work at these temperatures. When soldering copper to copper, or copper to silver with IT, mix about half-and-half Handy Flux and black flux (available from Indian Jewelry Supply or many welding supply stores). You can use straight black flux, but it is opaque when active and hard to see when the solder flows. If you keep heating to try to force it to flow after the seam is dirty, the copper and silver will fuse instead of becoming soldered. Once this fusing process has begun, it will continue each time the piece is heated and cause the silver to "disappear" by alloying with the copper. Enamel will also tend to pop off the alloy because the metal moves every time it is fired.

Any soldered joint will be subject to eventual break down in the enameling process. With IT solder the break down is very much slower than with hard solder. It is possible to successfully complete a piece with multiple firings using hard solder on unstressed joints, but care must be taken not to over-fire the piece, which will cause the solder to break down and endanger the enamel around it.

If I am attaching heavy cloisonné wire or other fairly large metal elements to the surface, I usually tack them in place. Large or heavy pieces are difficult to get to stay on a steep wall with only enamel holding them in place. After the soldering is completed and all the excess solder removed, the piece is pickled and then rinsed in water and baking soda to neutralize the pickle. This rinsing is essential for any piece of metal that has been in pickle otherwise the enamel will be bubbly and porous.

I mostly use 80 mesh, medium or hard, old Thompson leaded enamels, but the Japanese leaded enamels are wonderful. I order these when I need to replace old colors. I do not clean the enamels, but I do use different mesh sizes depending on the application, which generally takes care of cloudy transparents. I am careful not to apply transparents thickly, or if I need a thick coat, to sift out all but the larger grains. If I find some transparents that have been improperly stored and fire cloudy or foamy, I will sift and then wash them. To wet pack I use 3/0 or 4/0 brushes or a wire tool. The sifters I use I make with various meshes of stainless steel screening and plastic containers. I melt the plastic onto the screening with a warm to hot spatula. I find that the sifters you can buy break easily and are too shallow. For sifting on the enamels, I first spray on 1:1 diluted Klyr-Fyre with an airbrush and then use a variety of mesh sizes, mostly I use 80 mesh and 200 mesh enamels.

I begin enameling on the area that will be the hardest to reach for stoning because the enamel flattens a little each time it is fired. Usually this area is the inside of a vessel. If the vessel is a bottle shape, I pour in diluted 1:3 Klyr-Fyre, roll it around and dump out the excess. Then I pour in dry enamel, roll it around and pour out the excess. The piece is dried and fired to maturity. Sometimes I put a dab of enamel on the outside in order to tell when it is fired. The inside will take a little longer firing than the sample on the outside, so it needs a few extra seconds. If there are bare spots after this firing, then I wet pack over them and re-fire, repeating until the whole inside is covered.

If the piece is more open but the walls are fairly high and steep, I usually wet pack the inside with 80 mesh enamel and 1:4 diluted Klyr-Fyre. For this shape, I wet pack the enamel in 3" bands leaving a narrow bare line between each band. This line will prevent the weight of the enamel at the bottom from pulling the enamel nearer the top down into a pile at the bottom. Two or three coats are needed on the inside, so the bands of enamel are staggered if there is bare metal to cover. If the piece is open and shallow, then the Klyr-Fyre is diluted with water to 1:6 solution.

The outside is enameled by whatever method you want to use. After one or two complete coats have been fired on both sides, I trim the overhang of the rim on the side I enameled first. This prevents the enamel on that side from pooling around the rim and becoming too thick if most of the later firings are done with the piece upside down.

If I am going to embed cloisonné wire, I apply the base coat and fire it three times to harden it. After I cut and shape the wires and set the vessel horizontally on a firing support, I place the wires on that exposed section with uncut Klyr-Fyre or wetted Japanese lotus root powder, let it dry and then fire the wires into the enamel. The piece is rotated until all the wires are embedded. Any unwanted scars that remain from the firing points are later cleaned and enameled.

From this point on, enameling a vessel has about the same demands and possibilities as enameling a flat surface. The only difficult part is achieving a consistent underfired surface since the part of the vessel in the back of the kiln tends to fire faster. The enamel is stoned, the uncut overlap of the rim is trimmed, and a final finish is put on the metal just before the last firing.■

"Folded Vessel,"1998 vessel, 7" x 4.5" x 4/5", enameled on fine silver, raised, fabricated. *Photo by artist*

"Rosewater Itradan" vessel and cups, 7" x 12" x 3.5", raised, fabricated, enamel on fine silver, coral, ebony tray. *Photo by artist*

"#2114" vessel, 10" x 5.75" x 3.5", electroformed copper and enamel.

ENAMELING ON
ELECTROFORMED VESSELS

JUNE SCHWARCZ

June Schwarcz was introduced to enameling by a student of Kenneth Bates. She was immediately interested in the transparent qualities of the enamel. She is known for her enameled electroformed vessels. Her work is in many collections, too numerous to list. Fifty years later, she says, "I feel that I have not exhausted the possibilities of working in this medium."

June Schwarcz in her studio

Electroforming is electroplating when the plating forms most of the substance of the object. Plating thin copper foil until it is as thick as you want is considered electroforming. Information about electroplating is in *Metal Techniques for the Crafstman,* by Oppi Untracht. Most of my work is enameled with transparent enamel, sometimes using a little opalescent or opaque enamel. My pieces vary in size from 3" to 12" in height. I do not repeat my designs, although I do make variations of an idea. I also enamel in other ways.

I started to enamel in 1954 using spun copper shallow trays or flat copper sheet for wall pieces. Later I began to hammer most of my bowls. For my basse-taille pieces, I apply transparent enamels that can be used directly on copper. Most transparent reds, pinks, oranges, and so forth are not successful directly on the copper. I sift different colors in different areas. Some colors are applied more thickly than others. I then sift soft flux over the whole piece to even out the enamel so that none of it will be too thin and burn out during the first firing. On subsequent firings, I can use any color and build up the enamel according to aesthetic considerations.

In 1962 I became interested in electroplating to gain greater depth for my etched basse-taille pieces. My husband, Leroy Schwarcz, a mechanical engineer, brought home a sample of very thin foil. I think it was about 1 mm thick. It proved to be good for making three-dimensional pieces. It had something of the quality of soft fabric and could be made sturdy with electroplating

My shapes are developed with newsprint paper, which is then used as the pattern for cutting the foil. The foil can also be pleated or gathered before the plating. I use very thin copper wire to sew the foil together. Sometimes if the foil is too thin, it can become distorted in the plating bath. To prevent that, I paint the inside with melted wax. Heavier foil does not need the wax. It is important that the copper be absolutely clean just before putting the piece in the plating bath. I clean the copper with Sparex2 and rinse it thoroughly. After plating, the wax must be completely removed, and I burn it off. The copper piece is then cleaned again with Sparex2 before enameling.

I have a 30 gallon plating tank for my bath. The formula I use is 28 - 30 oz copper sulfate and 8 - 10 oz concentrated sulfuric acid per gallon of water. I have had better results using a prepared copper sulfate solution, although it is possible to plate with a solution made of agricultural grade copper sulfate. Over time, the amount of copper in the solution seems to increase in the solution. I test it and sometimes have to dump out some of the solution and add water and sulfuric acid to get it back to its original condition. I test it when the plating begins to get rough.

My kiln is 15.5"w x 18"d x 18"h on a 220V line. I fire at 1500°F. The number of times I fire depends upon aesthetic decisions. Sometimes the

"Adam's Pants #1, #2194" vessel, 11.5"h x 5" x 7", enamel on electroformed copper foil.
Photo by M. Lee Fatrherree

copper on the inside blisters, which is my greatest problem. Sometimes I enamel over the inside, but then sometimes bigger blisters break and get black. If that happens, I clean the piece and enamel over it, but the irregularity shows. Although sometimes I wet inlay the enamel, I usually sift my enamels onto the piece. The technique depends on the design. I use 80 mesh leaded enamels; some are old Thompson enamels and others are from other countries. I have found them all satisfactory, but some are more powdery than I like. The powdery ones I sift with Thompson's 200 and 300 mesh sifters to remove the fines.

The inside of the vessel is enameled first. I had some sifters made of plastic tubing or little bottles that were attached to dowel sticks of different lengths. The long handles help to sift into the vessels. I use an airbrush to spray a Klyr-Fyre solution 1:2. I spray and sift twice. It is important that all the particles of enamel have been moistened with the adhesive. An avalanche of enamel can occur in the kiln if the enamel is not adequately moistened and dried to stick on the piece before firing. The piece is placed in the kiln with the bottom toward the back elements. Another layer of enamel on the inside is often needed before I can work on the outside of the piece.

Occasionally, after the enameling is completed except for the final firing, I use a raku technique that I developed. The colors vary according to the metallic oxides in the enamel. I fill a large canning pot with fresh, green, dry Acanthus leaves. After the final firing, I take the piece out of the kiln and quickly put the pot of leaves upside down over the enamel piece. The piece has to be fairly large for this to work because a small piece will not stay hot long enough to burn the leaves. The pot should be left in place long enough for the piece to cool down. You can always fire the enamel again if you do not like the raku effect. ∎

"#2225" square vessel, 6.25"h x 9" x 9", enamel on electroformed copper. *Photo by M. Lee Fatrherree*

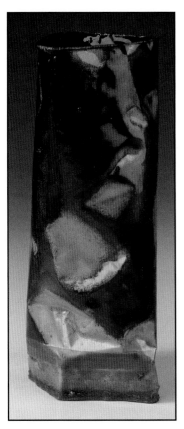

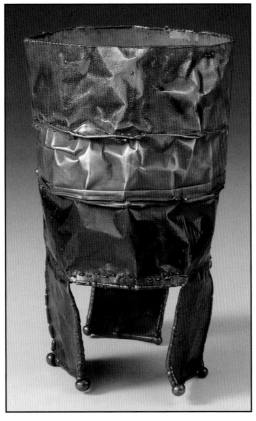

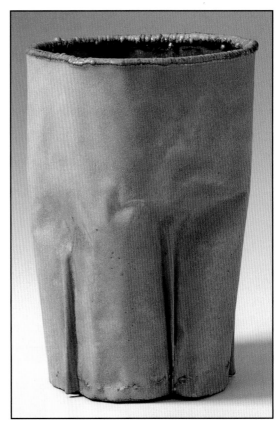

"#2219" vessel 10.5" H x 4" x 3", enamel on electroformed copper foil. *Photo by M. Lee Fatrherree*

"#2229" vessel, 8"h x 5.5" diameter, enamel on electroformed copper foil. *Photo by M. Lee Fatrherree*

"#2238" vessel, 5.5"h x 4" diameter, enamel on electroformed copper foil, sand blasted. *Photo by M. Lee Fatrherree*

PORTRAITS WITH ENAMEL WATERCOLORS

ORA K. KULLER

In her last year at Art College for teachers in Israel, Ora Kuller was introduced to enameling. In the United States, she studied enameling with Judy Danner and jewelry with Becky Brannon. In a workshop with Rebecca Laskin, she learned special enameling techniques. These teachers changed her life as an enamel artist. The magic and beauty of enameling captured her. She gives enameling workshops and works in her Belmont, MA, studio, developing her own way of portraying her thoughts and dreams through enameling.

Ora K. Kuller

"Tamar" plaque, 9" x 11", segments are 5" x 3" and smaller. 18 ga copper, fine silver foil, fine silver cloisonné wires, and enamel watercolors.

My design for a portrait in enamel is based on either a photocopy of a photograph or my sketch of the subject. The work is divided into small sections of various sizes and shapes. Each piece is an almost independent unit. Cutouts made from photocopies of the design are used for cutting the metal, blocking the area, transferring designs and shapes, etc.

18 ga copper or fine silver is used for parts that are fired many times. Less complicated sections are made of 20 ga metal and for tiny pieces that will be fired just three to four times, 22 ga or even 24 ga will suffice. I use a frame saw with a 2/0 blade to cut 18 ga metal and a 4/0 or 6/0 for thinner metal and very intricate designs. The metal is first covered with a layer of masking tape, then a layer of double-faced tape. The third layer is the pattern for cutting. After cutting the metal, the layers can be peeled off together and saved. Should a need arise, the pattern is ready to be placed on a fresh piece of metal and re-cut. Holes are drilled in the metal if it is to be connected with screws or other connecting elements. The metal must be cleaned before it is enameled. Annealing in the kiln cleans fine silver, but I heat the copper in a 1500°F kiln till it turns gray green and then clean it with Penny-Brite.

All of my enamels are from Thompson Enamel Inc. I purchase their unleaded 80 mesh enamels and use #533 White Liquid Form enamel as counter enamel on the back of all the segments. This liquid enamel should be the consistency of thin yogurt. It can be poured over the metal or applied with a soft brush. When the counter enamel is dry, I sift enamel on the front.

Sifting the enamel, without any adhesives, through different mesh sifters helps to create a perfect surface for painting. The first sifting of a pale color enamel, like #1202 Off White, is done through a 60 mesh sifter over the whole area of the head, followed by a thin layer of #1202 through a 100 mesh sifter. A last sifting of the #1202 through a 200 mesh sifter is over the whole area, with particular attention to the edges. The tiny grains from the 200 mesh cling better to the edges than the bigger grains and will prevent the edges from burning out. The piece is fired at 1500°F for about 1-1/2 minutes or until the enamel, inside the kiln, looks even and glossy. Removed from the kiln and cooled, the piece is stoned under water with 150 and 200 Alundum stones to ensure a flat surface. In painting a portrait, mounds and hills in the enamel can distort the expression of the face.

Sometimes I sift #1010 White only on the face and neck using a paper cutout to block the hair section. After removing the paper cutout, I paint the bare copper with Scalex leaving a narrow space between it and the sifted enamel. When the Scalex dries, the piece is fired until the enamel is shiny. When cool, the Scalex and the firescale are removed and the bare copper cleaned.

The hair section is painted with a very thin layer of oil base P-3 Black Underglaze and dried on top of the kiln. When dry, the pattern of the hair is sgraffitoed with a metal or wood scriber. Masking the white enamel of the face with the appropriate paper cutout, I sift 200 mesh granules of transparent enamel over the design of the hair and fire the piece until the enamel is shiny and transparent. P-3 Black Underglaze should always be fired with transparent enamel over it.

At this stage, I turn the piece over and sift over the fired counter enamel the same enamel that was sifted on the front. Using the same enamel on the front and on the back reduces the risk of cracks in the enamel.

The features of the face can be drawn free hand, but sometimes I use a copy of my drawing to transfer the outline of the facial features. I make a carbon paper that will be harmless to the enamel surface: Over the flame of a candle I hold a white china plate at an angle that will allow the flame to cover the plate with black soot. I wipe off the soot with a cotton ball and transfer the soot to the back of the portrait cutout pattern. The lines of the design can be traced by placing the blackened side of the drawing on top of the enamel. The transferred lines will appear on the enamel surface very faintly and will give a hint of the correct positions. The next step is painting with the watercolors.

"Serenity" wall piece, 11.25"h x 13.25"w, Limoges enamel , cloisonné, under-fired enamels on copper, fine silver foil, moonstone, diamonds.

"Love's First Moment" (detail), enamel wall piece, 10" square, Limoges underfired, cloisonné enamel on copper, 24k gold wires, fine silver sheet and wires.

"Sorrow, East Europe" enamel wall piece, 24"h x 34"w x 1"d. Limoges underfired, cloisonné, breaking through, liquid enamel, etching, fine silver sheet and foils, enamel on copper, garnets, amber. Wood frame painted with watercolors, velvet embroidered with silk yarn.

"Sorrow, East Europe," detail.

The watercolors come in powder form. The three kinds that I use are Ceramic Pigments, Overglaze Painting Color, and the 400 mesh fines of the enamels. Ceramic Pigments are not enamels and will not fire to perfection without an enamel addition of either PF-1 Painting Flux or 913E Mixing White. When mixed with Painting Flux, the colors retain their intensity whereas when mixed with Mixing White, they become lighter and softer. The ceramic pigment and the Mixing White or the Painting Flux are mixed 1 part pigment to 4 parts of either the white or the flux. As I mix the colors, I add one to two drops of Klyr-Fyre and a few drops of distilled water, just enough to achieve a good consistency for drawing and painting. The mixture should be smooth and silky because lumps do not fire well or fire with a different intensity of color. If these mixtures are kept dust free, they can be used for a few months. Since some of the enamel watercolors and pigments look very different before and after firing, I have one sampler of the colors mixed with the Painting Flux and another with the Mixing White.

With the outline of the features traced on the enamel, I go over the lines with the enamel watercolor. Dipping a 20/0 brush in the darkest color combination that I have, I render thin and delicate lines. The color I use is a mixture of dark brown and dark blue. Drawing with these enamels is not like working with regular watercolor paints. The enamel surface is non-porous, and if I cross over my painted lines, the lines underneath might disappear. The colors dry very quickly and become powdery, and so I have to be very careful not to drag lumps of powder with my brush. On the other hand, I can improve the drawing with ease. With a moist brush the lines can be refined until they are as thin as a hair. I can shape and reshape the drawing until I am content.

The piece is fired at 1450°F-1500°F for about one minute and 20 seconds, or until the lines become darker inside the kiln. Before firing again, a fair amount of work can be done. The shadows can be defined, dark values can be enhanced and the lighter areas can be softened. The addition of Klyr-Fyre to the color makes the paint adhere better to the enamel surface and more layers can be added without disturbing the unfired layers below. As enamel colors tend to be lighter before the firing and darker after, the shadows and lights over the face and neck may be too contrasting. Layers of Mixing White will soften the harsh shadows, and at the same time, if I want, I can add skin color to the portrait.

For a skin color over a base coat of #1010 White, I mix the tiniest amount of OC-70 Red, OC-32 Yellow, 1715P-Clover Pink and a larger amount of Mixing White. If the base color is Off White, I use browns for skin color, OC-82, OC-83, OC-85, a little 906E Green and OC-95 blue. The lips are drawn with OC-70 Red and OC-71 Orange. A hint of blue will enhance the white of the eyes. Often during the many firings, the lips will need another application of color. A portrait can require 20 or more firings. It is a slow procedure that requires patience.

During the intermediate steps of drawing and painting the portrait, I take care to fire the piece only imperfectly, otherwise by the fourth or fifth firing the lines and colors may have vanished or been distorted. Enamel watercolors are very delicate and do not tolerate high or prolonged firing. Only when I sense that the piece is nearly perfect will I paint the whole appropriate areas with a thin layer of the skin color mixture and fire a little longer. Delicate details can be added just before the piece is fired to perfection. I employ various techniques and colors to create the design that surrounds the portrait.

I usually connect the enameled pieces to a wood panel with Silicon II. I also use tiny nails that I make from fine silver or copper to hold the enameled pieces in place. I feel the nails echo and enhance the design. When all the pieces are glued to the board, these elements are threaded through the previously made holes to the supporting wood panel. They also add mechanical strength to the work. Finding or making a frame that will compliment the work without competing with it is the last step. ■

"Jerusalem's Menorah" free-standing enamel object, 9"h x 7"w x 2.25"d,
Limoges, liquid enamel, pierced silver, copper, sterling silver, fine silver foils.

PAINTING PORTRAITS WITH CERAMIC PIGMENTS

"Adam" plaque, 4"x 5", painted oxides and ceramic pigments on enameled copper plaque. *Photo by artist*

STELL SHEVIS

Stell Shevis has been a professional painter and printmaker with her husband, Shevis, since they graduated from Massachusetts College of Art in Boston in 1937. Her interest in enameling began when she inherited a huge supply of enameling equipment. Stell's enamels have been juried into shows throughout this country and abroad and also published several times in Glass on Metal.

Stell Shevis

I have been a painter all my life and I was delighted when I first learned how to paint portraits with ceramic pigments in a workshop with Bill Helwig. The technique is using a painting method for enameling. I work on either the pre-enameled steel tiles from Thompson, the heavier ones not the newer, very thin ones, or 16 or 18 ga copper. For size, I like something that I can hold easily in my hand, so usually 4" x 4" or 4" x 5" for portraits, although I recently did one that is 5" x 7". I often use a larger size when painting landscapes; the largest I've done is 10" x 10".

I counter enamel the tiles and fire with two coats of any good pale opaque. I like op.760 off white, op. 339 sandalwood and op. 852 opal beige; they are all 80 mesh Thompson leaded enamels. I inherited a huge supply of them, but you can use any of the lead free opaques that appeal to you. I do not wash them.

The ceramic pigments from Thompson , the OC series, are not an enamel. They are oxides of metals like copper, iron, manganese, and so forth, and must be used over or under a vitrifiable glaze. Woody Carpenter of Thompson Enamels answered my questions about this material, since I wondered why they are called ceramic pigments. He said the word ceramic comes from the Greek word *keramos*, which means "burnt stuff." Ceramicists (potters) use these coloring compounds in their stains and glazes. Now you know as much as I do!

A tiny bit of these colors goes a long way. I use one of the white enameled tiles as a palette and place on it a bit of color the size of a split pea; I use an old, narrow flexible painting knife. Put a drop of squeegee oil next to the color on the palette and mix some of it thoroughly with the knife. This mixture should be fairly stiff, a little stiffer than toothpaste. Add two or three drops of #5 thinning oil or oil of lavender beside it on the palette; I keep an eyedropper handy for the purpose or put your oils into those little bottles with dropper tops. Mix the thinner oil with the first thick mixture until it is thin enough for painting. Using small watercolor brushes, #00 or #1, do some experimenting with various brush strokes to learn how thin a mix you need to get the effect you want. I use it pretty much as I do water color. My favorites in the series are OC 85 red brown, OC 16 blue, OC 191 green, OC 50 black, and OC 32 yellow. I prefer these because they will fire without any change in the color. The finished piece after firing will look the same as it does before firing, although some of the oranges and reds do change if they are not fired very, very carefully.

"Lew" plaque, 4" x 5". painted oxides on enameled copper plaque. *Photo by artist*

I do not make a preliminary drawing on paper, but work directly from the model. The person will sit a few feet away in front of me in a comfortable chair. All my colors, palette and tools are on a small table at my side. I hold the tile for a minute, rubbing it with my hands to warm it slightly and add any personal oil from my hands. This preliminary routine seems to be a friendly way to start.

I begin by using my #1 brush dipped into my mix of OC Brown to sketch the placement of the head. Any mistakes are easily removed with a bit of tissue or paper towel. I keep a pile of cut paper squares about 3" x 3" ready for wiping out or making corrections. After placing the head, I dab in the shadow areas and draw the features, still using the red brown. The OC-85 brown mixed with OC-16 blue makes a pleasing dark for hair or eyes. It is a good idea to keep separate brushes for each color in order not to have to keep cleaning the brush. Each artist will find a favorite way of working, it just takes practice.

A stiff bristle brush pulled through when the color is slightly dry gives a nice feeling of hair. All of these colors may be mixed to get various shades. Usually one firing at 1450°F for about 90 seconds is enough. The piece should look glossy all over. If you take the piece out of the kiln before it's glossy, put it back for a little longer. As every enamelist knows, each kiln is different. If fired too high or too long, the color sinks into the enamel and leaves depressions. It's important to make some test pieces before going on to a serious piece in order to become familiar with the way your kiln fires.

Sometimes a second firing is needed if you wish to strengthen an area or add a background. In the portrait of Cary, I added the pink background in a second firing. For this pink I used a very finely ground enamel mixed with the oils in the same way; it is an old one with no brand name but has #431 on the label and is probably leaded. I use this pink sometimes for color on cheeks or lips because I can't seem to get a good pink with the ceramic pigments. Lilyan Bachrach uses overglazes from Standard Ceramic Co., especially for pinks and violets, and china paints from other companies.

You will find that the thinnest veil of color works best. If you make it too thick on the surface, the color will not be completely absorbed into the underlying enamel after firing and that area will look rough and dull compared to the gloss of the rest. A very thin sifting of a soft flux will heal it. I have learned to handle the color so that I can usually finish the piece in one firing, two at most. I do not sift flux over the piece after the first firing because if the painting is done right, the color is absorbed into the enamel, and a pleasant gloss is over all.

It takes some practice to find your own way of using these pigments, but if you enjoy painting, you'll love this technique.

Tools and supplies needed:
· Enameled steel or copper tiles to base coat, any size
· Palette and small flexible painting knife
· Good quality small watercolor brushes #000, 00, 0, 01, and a few larger
· A stiff bristle brush, 1/2" wide, such as used for oil painting
· Paper towel, tissue or lint free rags for corrections
· Small bits of sponge, smooth leather, rubber eraser or rubber tip tools
· Plastic wrap for covering the palette when not in use.
· Ceramic pigments: Thompson C 85 brown, 50 black, 16 blue, 169 Yellow, 191 green.
· Squeegee oil, #5 thinning oil, A11 oil of lavender (try them all)
· 80 mesh opaque enamel colors of your choice, lead free or leaded■

"Jean" plaque, 4" x 5", painted ceramic pigments on pre-enameled steel plaque. *Photo by artist*

"Cary with Pink" plaque, 4" x 5", painted ceramic pigments on pre-enameled steel plaque. *Photo by artist*

"Peter" plaque, 4" x 5", painted ceramic pigments on pre-enameled steel plaque. *Photo by artist*

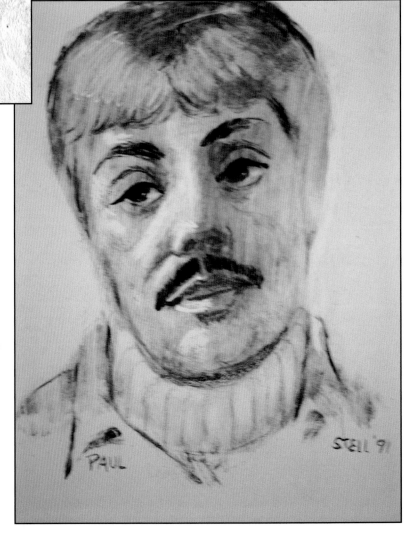

"Paul" plaque, 4" x 5", painted ceramic pigments on pre-enameled steel plaque. *Photo by artist*

JEWELRY-SIZE MINIATURE PAINTINGS

MONA SZABADOS

Mona Szabados was born in Oslo, Norway. She has been enameling since 1980 and is mostly self-taught. Her goldsmith husband, Alex Szabados, makes all of the gold and stone settings for her enamels.

Mona Szabados.

My enamel jewelry includes earrings, pins, pendants, lockets, and rings done in grisaille and Limoges techniques. Within most of the pieces are miniature paintings with images of women's faces and animals. My enamel pieces, using copper, 80 mesh enamels, ceramic pigments and foils, usually require from 25 to 40 firings. I collaborate with my husband, Alex, who is a goldsmith. Each enamel piece is set in 22k or 18k gold.

Usually I start my enamels by cutting out a copper shape and then do a rough sketch of the enamel for that piece. Sometimes I do the sketch first. Most often I have a very well defined color arrangement in mind, but it can change as I work.

I order 18 ga and 20 ga oxygen free, hardened copper (OFHC) in sheet form from American Copper & Brass in Oakland, CA. I use the Japanese 246 gold foil and pure silver foil from Thompson; I also use palladium leaf from Enamelwork Supply Co. The kiln I use the most is 9" x 8" x 6-1/2" on a 110V line, but I also have a large kiln on a 220V line. Firing is at 1350°F-1400°F.

After the copper shape is cut, the metal is cleaned by firing at 1300°F for five minutes, then put into the pickle of 30 percent Sparex2 and 70 percent distilled water, rinsed and then glass brushed. I clean anywhere from one to six pieces at a time, depending on my work schedule. My enamels are the leaded, standard 80 mesh. I wash one to two teaspoons of a color with distilled water until the water is clear. The washed enamels are stored wet in fish and tackle containers in a closed wooden cabinet. They are re-rinsed with distilled water before I use them. I do not make color samples; I learn about the enamels as I use them.

Using 1/0 to 4/0 1/4" hair, sable brushes, I wet pack, with distilled water, medium or hard flux and transparent enamels on the front. Flux is wet packed over the face area and dark transparents over the other areas, without the flux and the dark transparents touching. This firing leaves a firescale line between the areas of the flux and the dark enamels. The enamel is packed about 1mm high, which after firing barely covers the copper. About four coats are applied on the front and fired before sifting dry enamel on the back. The enamel piece is supported on a trivet that is on a mesh planche for the firing. I brush off the loose firescale on the back of the piece between the firings and also clean the edge of the piece with a carborundum stone. When there is an adequate base coat on the front, I sift counter enamel on the back over the remaining firescale. Any additional coats needed later on the back are wet packed.

Then I proceed with the firings on the front. Colorful transparents are added to the face for shadows and features. After each firing, the edge of the copper is rubbed with Carborundum stone to remove any firescale. For the face, I use Schauer's leaded opaque ivory color or white opalescent enamel, applying and firing about ten thin layers in the manner of grisaille, wet packing very wet and sculpting the enamel thicker on the forehead, the tip of the nose, etc. The thin areas of the opaque

"Locket of the Four Seasons, Summer" locket, 2.25" x 1.5" x .75", Limoges enamel on copper, approximately 40 firings, transparent and opalescent enamels, 24k gold wires, fine silver foils and granules, setting: fabricated 18k gold, 22k gold bezels, opal, diamond, sapphire.

allow the transparent enamels to show through. Schauer is no longer manufacturing enamels, but fortunately I purchased a large supply. If I am planning to use silver foil in an area, then I use a cool color enamel under it.

To cut the foil, I place it between two sheets of typewriter paper. After separating the foil and the paper, I pick up a piece of foil with a brush that is wet with a 1:1 solution of Klyr-Fyre and distilled water and place it on the fired enamel. I used to always put silver flux over silver foil before applying a transparent pink or red, but now there are some leaded Japanese enamels that fire well without the flux under them. The transparent enamels generally will stay a lighter color if a coat of flux is fired between the layers.

In the last three firings, I very carefully use Thompson ceramic pigments in very small amounts (tiny dots) to strengthen the features of the face. Each color is mixed with distilled water in a section of a party ice cube tray; and then a small amount of each color is placed on an agate slate and mixed with a little imitation lavender oil. I use the imitation lavender oil because it dries faster. A box lid covers the slate palette to keep the pigments clean between working sessions. After two weeks of work, or when I do a show, the slate is cleaned off to be ready for a clean supply.

When the enameling is completed, I smooth the edges with a blue wheel on the flexible shaft. I set the enamel in the bezel Alex has made, and he sets any stones we have chosen for the piece we have designed together.■

"Locket of the Four Seasons, Fall and Winter" locket. 2.25" x 2.75" x .375", Limoges enamel on copper, 24k gold foil and granules, fine silver foil. setting: hand-fabricated 18k gold, 22k gold bezel, opal, diamond, sapphire.

"Triptych" (open), 2.5" x 2.5" x .25", Limoges transparent and opalescent enamels on copper, 24k gold foil and granules, fine silver foil, approximately 35 layers and firings, setting: hand-fabricated 18k and 22k gold, gold granulation, opal and tanzanite.

"Triptych" (closed), 2" x 1.25" x .375", Limoges transparent and opalescent enamels on copper, 24k gold foil and granules, 35 layers and firings, setting: hand-fabricated 18k and 22k gold, gold granulation, opal, tanzanite. The doors open.

Necklace, Limoges transparent and opalescent enamels on copper, 24k gold foil and granules, fine silver foil, approximately 30 firings each, setting: hand-fabricated 18k and 22k gold, gold granulation, opal, diamonds.

"Spring" double-sided pendant with swivel, transparent and opalescent enamels on copper with 24k gold foil and granules, pure silver foil, approximately 40 layers and firings, setting: 22k gold with 18k handmade swivel. Opals, diamonds, sapphires.

OTHER MATERIALS FOR ENAMELING

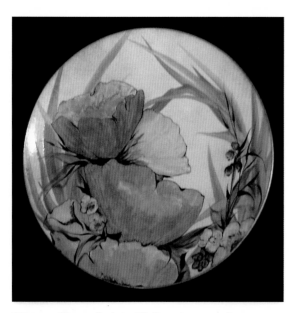

"Blue Flowers" plate, 9"diameter, overglaze painting on vitreous enamel base-coated copper, fine line black drawing. *Photo by Aaron J. Plotkin. Private collection*

OVERGLAZE DIRECT PAINTING

LILYAN BACHRACH

Lilyan Bachrach

Lilyan Bachrach, the author of this book, fell in love with enameling after taking a course in 1955 with Doris Hall. At the Worcester (Massachusetts) Center for Crafts, she studied silversmithing, jewelry, pottery, woodworking, photography, color, and design. In the early 1960s, she studied cloisonné with Joseph Trippetti. She graduated from the Worcester Art Museum School as a Fine Arts major in 1968. In addition to enameling commissions, she has sold her Bachrach Art Enamels since the 1970s, at the ACC Northeast Craft Fairs, galleries, and shops across the country.

For more than twenty years, I have been using over-glazes as a direct painting technique. My canvas usually is a fired, flux, and white base-coated enameled copper piece. Although I sometimes make a rough sketch for placement, I more often paint directly in an impressionistic style. Each painting is a new and different delight. Even after all these years, the firing process is still able to produce a surprise often enough to hold my interest when I watch the enamel change color as it cools. When the result is not acceptable or what I intended, I enjoy working my way out of a color or design error.

I make wall pieces, plates, bowls, mezuzahs, and switch plates. The switch plates and the mezuzahs are small enamels that I use to experiment with new color combinations or a new design. In my studio, my work has ranged from 1-1/2" cloisonné jewelry to sectional pieces comprised of 12" square segments. I use ceramic overglazes and china paints (on-glaze enamels) and also mix them together to make additional colors. My over-glazes are old ones, probably leaded, from Thompson Enamel and Standard Ceramic Supply Co. The china paints are from various china supply houses. Most china painters and enamelists mix the material with oils, but I use water to prepare the overglazes for painting. In the early 1970s, I produced a series of enamels made with fine line black pen drawings and over-glaze colors added with small brushes. While traveling in Canada I visited an enamel studio that had been marketing a line with the same technique. They, too, were painting flowers with the fine line black first. Shortly after that visit, I thought of using watercolor brushes first to paint the flowers and then adding the black pen line. The only overglazes I had were the limited colors that Thompson Enamel then carried. I stopped using this technique because I did not like the yel-

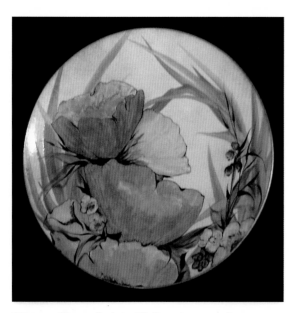

"Orange Flowers" plate, 7" diameter, overglaze painting on vitreous enamel base-coated copper, fine line black drawing. *Photo by Aaron J. Plotkin. Private collection*

lows and oranges when fired at 1500°F, and I thought the Thompson palette was inadequate. About a year later, while exhibiting at the American Craft Council Northeast Fair, I was browsing through the catalog of Standard Ceramic Supply Corp. and spotted columns of ceramic overglazes with enticing names from cream to dark purple. They fired at just under 1500°F, which was what I wanted. The sales representative told me they would not work on enamels and I would be wasting my money. I decided to try them anyway, despite his protests, and bought about 12 colors: yellows, oranges, pinks, purples and a deep blue. Most of them worked, and I am still using the Standard Ceramic Supply product. The ones I list below are my favorites. It was not until the late 1980s that I added the china paints.

If you want to add overglazes to your supply of enamels, buy the smallest quantity of about 8-10 colors to start, or a sample kit if it is available. A teaspoon amount lasts a long time. The china paints come in a glass vial. The violets and pinks are the most expensive. I am assuming that you know how to paint in some other medium. If not, you can use them to add an accent of color or shading to an area of fired enamel on your piece made with one of the beginner's techniques.

The Standard Ceramic overglaze colors I use most are: Lemon ST114, Pink 236, Green 112-P, Canary 650/291, Dark Violet 324 and Orange 286. I use the Thompson 900 series, except the yellow and orange. Originally, I made color samples on 3" x 8" 20 ga copper that had a medium-firing white base coat and black crackle on the back. My enamels are mostly 80 mesh leaded ones from Thompson that I purchased in quantity years ago. For the base coats I usually use Thompson's leaded 80 mesh #1005 medium flux, #1000 medium white, counter enamel, #124A hard black, #772 liquid form enamel and #426 soft flux over my name on the back.

I primarily use the #169 Norman kiln on a 220V line. The inside chamber is 16" x 16" x 9". The door opens horizontally from the right side. The pyrometer is set in the right rear corner. It has two variable control switches although I would prefer one switch. It was re-bricked and rewired about 15 years ago. I have a reserve set of wires.

To prepare the overglazes for painting, I put about 1/2 teaspoon of each color in the bottom part of a 3" plastic petri dish near one side. The cover is labeled. I add water gradually with a syringe as I blend the overglaze to a painting consistency with a small bent painter's palette knife. My favorite knife has a 1/2" wide straight edge because I accidentally broke off the tip years ago. I use tap water. When I paint, I tilt the uncovered petri dish slightly so the overglaze is at the top of the dish and a little water at the bottom of the dish.

I work in a modified production style for the base coats on purchased 18 ga copper forms. If I am making 6" to 8" plates, I usually prepare about 12 pieces at a time for the paintings, which then are painted in a number of sessions. If the base coat is to be a white or light opaque, I clean the copper pieces with Penny-Brite. If the base coat is to be a transparent, I clean the copper on the polishing wheel with a goblet buff that has been charged with Lea Compound C. When clean, each piece is placed back side up on a 2" bottle to be coated with black crackle, as shown on page 22. I brush on the crackle, let it dry and then sgraffito my name in the center with a sharpened chopstick. The loosened dry crackle is tapped off, and any dry crackle on the edge of the piece is removed with an edge of the square handle of the chopstick. The pieces are placed face side down on a tray and transferred to the enamel worktable. I heat the kiln to 1500°F and keep it there while I apply the enamel to two plate fronts.

Before applying the enamel, I set up the kiln furniture to receive two pieces for firing. On each of two 6" square nichrome mesh

"Bright Orange Flower" plate, 11.25" diameter, overglaze painting on vitreous enamel base-coated copper, fine line black drawing. *Photo by Aaron J. Plotkin. Private collection*

"White Flowers" plate, 9"diameter, overglaze painting on vitreous enamel base-coated copper, fine line black drawing, high fired. *Photo and Collection of Anina Bachrach*

"Abstract Bowl #3" bowl, 4.5" x 5" diameter, sifting and flowing opaques on flux enamel base coat, spun copper bowl.

planches I place a 3-pointed stilt for firing the piece with the front side up. While one piece fires, I apply the enamel on the next one, keeping a production line moving. This part is just work. By firing a number of pieces, one after the other, the firing hardware stays hot and prevents a big drop in the kiln temperature when the kiln door is opened.

For my usual white base coat on the front, I mix about 2 pounds of Thompson's leaded 80 mesh, #1000 white and 1 pound of their #1005 medium fusing flux in a 5 pound jar. I like the softer look with a little of the flux showing through instead of a bathtub white porcelain look. To sift enamel onto a piece, I use the 60 mesh, 2" old-fashion metal tea strainers that have sloping straight sides with the mesh flat across the bottom. Some of my sifters are labeled flux, white or counter. I only use them for what they are marked so I do not have to think about cleaning them. (See section on sifting base coats, page **XX**.) On the front, I sift a heavy coat of the flux/white mixture and a second light coat of white. The piece is sprayed with water before and after each sifting. With a light spraying of water, just enough to hold the enamel, I do not have to wait for the piece to dry.

The first piece is placed on the set-up stilt, put in the kiln and fired to maturity to ensure that the black crackle on the back is shiny and fired enough to adhere. While that piece fires, the next flux and white layers are sifted on the front of the next piece. The next piece is set on the other firing set-up, the first piece is removed from the kiln, and the next piece is put in the kiln. The program continues until all the pieces have a fired base coat on the front and on the back. This is the base coat method of firing both enameled sides in the first firing that I learned from Doris Hall in 1955.

My counter enamel, with the 60 mesh sifter, for the second coat on the back is 2/3 left over 80 mesh enamels and 1/3 Thompson's 80 mesh, leaded, 124A hard black. By having the hard black in the mixture, I do not have to re-enamel the back again. Soft flux, Thompson's #426, is sifted over my name. I sift two coats of counter on the back except over my name. As usual, a light spray of water is applied before and after each sifting. The same production line system is used, but with hammocks to support the plates, with the back side up, in the kiln. Each piece is fired to maturity. When removed from the kiln, the plate is transferred to a steel plate, back side up and weighted with an old iron until it cools. After all the pieces are fired, and loose firescale on the edges removed, the plates are ready to be painted. Each piece takes about seven firings.

I have a separate table for painting. The petri dishes, with the covers on them, are set up as a color palette. Any over-glazes that have dried out have a few drops of water added and are blended smooth to prepare for painting. Two bottles containing water are also on the table,

Three pins. Left: 2.5" x 2.5", cloisonné on repoussé fine silver, fused fine silver wire, setting: fabricated sterling silver.
Center: 2.5" x 2" cloisonné on planished 28 ga copper, 24k gold cloisonné wire, setting: fabricated gold plated sterling silver.
Right: 2" x 2.5", cloisonné on 28 ga fine silver, fused cloisonné wire, setting: fabricated sterling silver. *Photo by J.A.Perry*

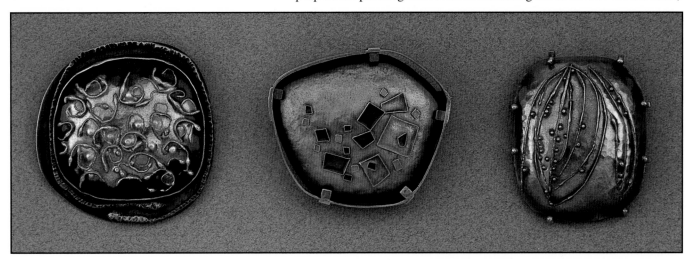

one for cleaning the brushes and the other for clean water, along with a folded paper towel for wiping a brush and the syringe and water spray bottle. I use sable or kolinsky watercolor brushes. As with any painting medium, you need to practice the brush strokes and develop your own style and preferences for colors. For painting leaves, I use Lebenzon's custom made, kolinsky, broad, long-pointed watercolor brush in the Chinese painting method. The brush is held vertically, pressing down on the heel of the brush for a broad stroke and gradually, as the brush is moved forward, pulled up off the paper for a pointed tip to the leaf. You need to remember that most china paints have lead in them and some have cadmium so do not put the brush in your mouth.

The overglazes handle like watercolor paints in that one wet color placed on another one will blend or bleed; but unlike painting on paper, the enameled surface does not absorb the overglaze. If part of the painting dries before you are ready to fire the piece, then the whole piece will need to be sprayed lightly with water. If the painting dries in sections, a line often appears between these sections when the piece is fired. The overglaze painting is dry by the time it is placed on the warm firing set-up and put in the kiln. Plates and bowls are placed in the kiln on a three-pointed stilt on the planche; plaques are fired within a hammock. I fire with my kiln at about 1500°F. It is years since I tested the pyrometer reading. I really judge the heat by the color in the kiln and fire for time by instinct and a quick peek. Overfired overglazes will lose the intensity of their colors.

For my floral painting, I often start by painting various shades of green leaves with the kolinsky watercolor brush. Next come my imaginary flowers. The first flower is often the predominating color from which the painting develops. If I do not like a part of the painting that has dried, I remove it with a small, stiff stencil brush. If the whole composition is not to my liking, the piece is rinsed off under running water. When the painting is at the point that it cannot be developed further, it is fired to just before maturity. Then if there are areas, say on the fired green leaves, where I want to add another flower, I either sift leaded 80 mesh #644 soft white in a shape and size or I wet pack an opaque over a leaf. To add color to this unfired white flower, I charge a brush with color and run the tip around the outline of the shape I want. A color can also be added within the added shape. The wetter the overglaze and the enamel, the more the overglaze color will bleed into the opaque white enamel. Instead of using the stiff stencil brush to remove part of the painting, you can use a wet clean brush.

"Orange Flower" plate, 6" diameter, 18 ga copper, over-glaze painting, fine black line. *Photo by J. A. Perry*

Footed bowl, 8" x 10", raised 18 ga copper, pewter rim, wet stencil. The design on the outside mirrors the design on the inside. *Photo by J. A. Perry*

Instead of mixing a color on your palette you can blend colors on the painted section when it is wet by charging the brush with thin overglaze and letting it run either where you direct it or by tilting the piece. Most of the overglazes are transparent, but adding the white will lighten the color and also lose most of the transparency. As you work with the overglazes you learn which ones need to be applied thicker, like the reds that burn out faster than other colors.

When I consider the painting completed, I often delineate some of the shapes by drawing with Hunt's #101 nib in a pen holder (you can also use a ruling pen) and Thompson's fine line black, which I purchase in liquid form. Some enamelists use it to sign their name on the front of their work. I call it "ink." I order it in the 1 ounce bottle because it takes a lot of stirring to put the glob at the bottom into solution. I use a dental spatula to stir the inkwell, and separate it into four little glass bottles with screw on lids. Before using the ink, it needs to be stirred well again. I first test if it flows from the nib like ink and test if I can draw some lines on smooth paper. I hold the pen almost straight up when I write with it. If the ink is too thick, add artist turpentine one drop at a time and stir after each drop. You will need to stir the solution about every ten minutes as you use it and also dip the pen in turpentine from time to time for the ink sometimes dries on the nib tip. After you dip the pen in turp you need to re-dip the nib into the fine line black a few times before you draw with it or the ink will be too thin. If you get an unwanted blob, wait until it dries and then remove it or part of it with a pointed chopstick. When the pen drawing is complete and dry, the fine line black is corrected with the chopstick. Then fire just until the drawing is smooth. You can test it, out of the kiln, with the edge of the potter's spatula. The wide areas will have break lines in them. I like that. If you over-fire, the drawn line becomes thinner and, of course, the overglaze colors are changed. I finish each piece with a sifted overall veil of 80 mesh soft flux.

The final step is to smooth the exposed metal edge. My sanding and polishing wheels are connected to a dust collector purchased and installed by a local dental supply company. Even so, I don a nose mask, a facemask, leather gloves, and a shower cap over my hair before using the equipment.

The face edge of the piece is finished first at the belt sander with a 6" x 48" fine grit emery cloth belt. After grinding the front and back edge of each piece, I remove the grinding marks. I tear a sheet of fine emery cloth into 1" x 12" strips and wrap one strip around one end of a 1" wide wood stick. Wearing leather gloves, I rub with the strip of emery cloth across the grinding marks to obliterate them. As the beginning wrap of the emery cloth strip wears out, I wind it a little to have a clean area to work with. The final finishing is a rubbing of the metal with a wad of 00 steel wool. I place felt tabs on the back of the piece over the three stilt marks and then add my label. Each piece is placed in a plastic bag to keep it clean.

Another use I have found for the overglazes is to make pale opaque enamel colors when I need to match a client's color swatch. For a small amount of enamel, about an ounce of 80 mesh, maybe white, is put into a 4 ounce glass jar with about 2 ounces of water and 1/4 teaspoon of over-glaze color in painting consistency. This mixture is then well stirred and allowed to sit for about 15 minutes and stirred again. If you want it darker, gradually add more overglaze, stir and let it settle in. Pour off the excess water. Place the open jar with a loose piece of paper on top of it on top of the kiln to dry out. When the enamel is dry, you can wet pack or sift as you wish. If I am aiming for a specific match to a swatch, I often have to make a few different batches and fire samples on scrap copper.

If you like to paint, you will find the nuances of color that can be obtained with overglazes limitless. The tactile quality of enamels as well as the sense of their enduring quality will bring you delight.■

STENCILS AND WATERCOLORS

JENNY GORE

Jenny Gore, of Adelaide, South Australia, was trained in graphic design and is a self-taught enameller. She started enameling in 1973. She works mainly on wall pieces and some jewelry. She has exhibited and/or given workshops in Europe, England, the United States, India, Japan, Mexico, Korea and Australia.

Jenny Gore. *Photo by Trevor Fox Photography*

One of my favorite techniques for achieving almost any kind of imagery is sifting dry enamels with cut or torn stencils. Over almost 30 years of enameling, I have developed several different methods for specific purposes and combining many techniques into one enamel piece. It is part of the challenge.

When working on commissions, and as I did in my early work, I begin the process by designing on paper and preparing a working drawing. Unless the enamel is to be a very large one, the drawing is true to size. This drawing is used as the pattern for cutting the stencils. For the stencils, I use acetate or drafting film, which will not disintegrate or stretch out of shape when it is wet. Areas of the drawing are colored with pencils to make the process easier, using one color and one layer of film for each layer of enamel. I then use the same color of waterproof marking pen to trace the areas to be cut. If necessary, some registration marks are made, keeping in mind the order of firing, e.g., the higher firing colors first and the softer ones near the final coats of enamel. Sometimes more than one color can be used in the same layer if the shapes do not touch each other.

On the drawing I number each piece, both the positive and the negative, and transfer the same numbers to the corresponding layer of film. For a complicated design, all the stencils marked in the same color should be kept in separate containers as they will be used several times, depending on the number of firings.

For some applications I prepare a collage using cut or torn paper from magazines. I arrange them on a paper of the same color as my base coat, and when satisfied with my design, paste them into position. If these are to be used to show a client, I mount them on a board and cover with transparent shiny film. I usually make a test piece in enamel also as the papers are seldom in colors found in enamel. This is a good opportunity to experiment with color mixing, especially with opaques, taking care to test them together before using them on the piece.

In recent work, I have allowed the design to "happen." I cut or tear stencils and build up layers of colors and shapes intuitively. This method is very much more fun than carefully planning everything and often leads to quite unexpected imagery. My individual enamel pieces range from 2" square up to 16" square. However, I can build up units of these to any size. I have several kilns. The largest is 17" x 17" x 12". The one I use the most is the 11" x 11" x 4". The small one, 6" x 6" x 4", I use only for test pieces. Here in Australia we have 240 volts.

I use l mm (about 18 ga) copper as a base for my wall pieces and 1.2 mm (about 17 ga) thickness for the larger pieces. The copper is cleaned with Amway metal cleaner and rinsed before sifting 80 mesh counter enamel on the back. The piece is fired. Unless I want to use the firescale as a part of the design, it is removed with diluted nitric acid, followed by a detergent wash.

"The River Is Wide…The Path Is Near," 13.5" x 13.5", enamel on copper, 24k gold and fine silver foils, 24k gold leaf fused to surface. *Photo by Trevor Fox Photography*

"Lazy Days," wall panel (Wright House commission), 13.5" square, sifting with stencils, enamel on copper, gold foil, gold leaf. *Photo by Trevor Fox Photography*

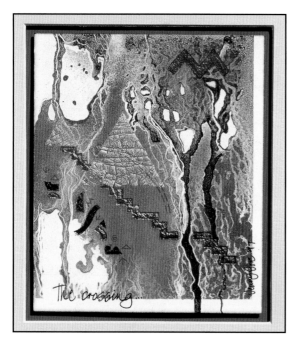

"The Crossing" wall panel, 4.75" x 5.75", painting and watercolor enamels on copper, gold and silver foils, gold leaf. *Photo by Trevor Fox Photography*

For the front, I sift on two coats of medium fusing white and fire the piece. I stone the edges with carborundum stone to remove the scale and any enamel from the edges and clean with detergent between all firings. Then the first layer of color can be applied. I like working on a base coat of either white or black. I use the medium fusing for the base coat for I find that a harder enamel sometimes causes craze or stress lines in the top coats.

To use the stencil, I spray the piece lightly with diluted Klyr-Fyre, position the first stencil and press it flat. I then spray a mist of the diluted gum over the stencil also using a spray gun and an air compressor, although aerosol spray packs or pump action spray bottles will achieve the same result. As I sift enamel over an area, I spray it lightly to hold the enamel in place, especially if the area is to be sgraffitoed. The piece is placed on top of the kiln to dry. The kiln temperature ranges from 800°C (1470°F) to 850°C (1560°F) degrees for the early layers and drops back to 750°C (1380°F) to 800°C (1470°F) near the end, for two to three minutes depending on the size of the enamel piece.

Further layers of color and shape follow, with any drawing by sgraffito or textural effects included along the way. If it is necessary to reuse a stencil, it can be scraped clean of enamel and rinsed to avoid contaminating colors.

When all layers are completed, some areas of fine silver or 24k gold foil may be fused to the surface and covered with well-washed transparent enamels by dry sifting or wet inlay. Any writing or drawing may be done with a pen using ceramic oxides or Carefree Luster. Sometimes these pieces have as many as 30 firings and along the way they tend to "grow," which means that stencils often have to be re-cut to fit precisely. With this technique, extremely complicated imagery with precise edges and fine detail is possible.

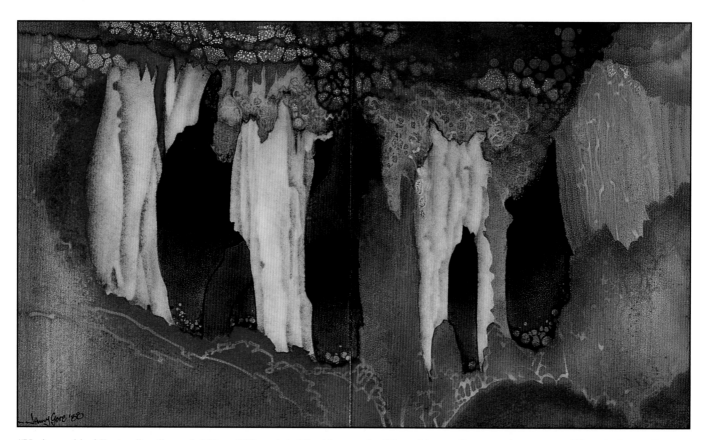

"Underworld of Fantasy" wall panel, 18"w x 12"h, painted liquid enamel, sifting with stencils, enamel on copper. *Photo by Jenny Gore*

WATERCOLOR TECHNIQUE

The copper base is prepared as for stencils, but the ground coats of enamel on the front are preferably white or a light opaque color in a medium fusing enamel. This fusing quality is necessary because most painting enamels are very finely ground and fire at a lower temperature: they are susceptible, therefore, to burning out easily if overfired. I apply and fire at least two thin coats for the front base. If I want a textured or watercolor paper effect, I leave the second coat at the orange peel stage, slightly underfired.

Thompson's pans of watercolors supplied in a waxy base are used either alone or in conjunction with other painting or screening enamel powder. I mix the watercolors with water and do not mind if they do not totally dissolve, as this gives a grainy look on the textured background. I first mix enough water in each pan of color to make either the painting or pouring consistency needed. You only need small quantities, and the pans last for ages. I use them on very small pieces (100 x 100 mm square—about 4" square). If I am using the fine powdered painting enamels, I stir them first with water, and then mix them on a piece of glass or on a plate with a palette knife.

When the application is completely dry, it can be worked on with sgraffito tools, and any undesirable areas can be removed. After firing at about 750°C (1380°F) for about 2 minutes, other layers can be added as desired. Transparent or opaque colors can be sifted on to complete the imagery in addition to accents of foils, leaf, writing or drawing with ceramic oxide or Carefree Lusters.

As you have probably deduced, my method of working comes from many years of firing enamels and mostly knowing what will happen in the firing process. It is the little surprises that make this an intriguing medium. ■

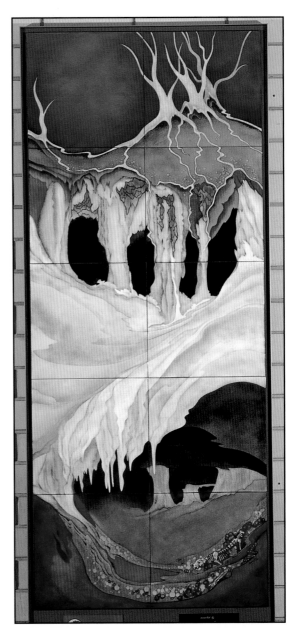

"Beneath the Surface" wall panel, 30"w x 75"h, painted liquid enamel, sifting with stencils, enamel on copper. *Photo by Trevor Fox Photography*

"Annagare, A Summer Dreaming" wall panel, 16.5"w x 20"h, grisaille, sifting with stencils, enamel on copper. *Photo by Jenny Gore*

DECALS AND CERAMIC PENCILS

JUNE E. JASEN

June Jasen began enameling full time in 1979. She has participated in numerous national and international exhibitions, competitions and received many awards. June has given workshops and lectures in the United States and in Europe. Her work is included in several archives, corporate and private collections.

June Jasen

A decal is a design or picture printed on specially prepared paper for transferring an image to glass, wood and other materials. Developed in the 1700s for the ceramic industry, they were used as an easier and less expensive way to produce and imitate china painting.

The patterns, usually silk-screened, are made into decals with organic oxides, glazes or stains with an inorganic fixative. Books are available that tell you how to make them. I prefer to buy the decals, as it is easier and more cost effective for me to shop at a ceramic hobby and supply store. Hobby ceramic trade shows have vendors who sell them. Factories that produce decals for the porcelain, enamel and glass industries will make the decals you design for a large order.

I alter a decal that I buy to make it mine. Starting with a large decal, I cut it up to suit the scale and aesthetic needs of the final enameled piece. For years, I made enamel dresses, which needed a pattern, or a bridal gown, which needed a bridal bouquet. Decals were a perfect solu-

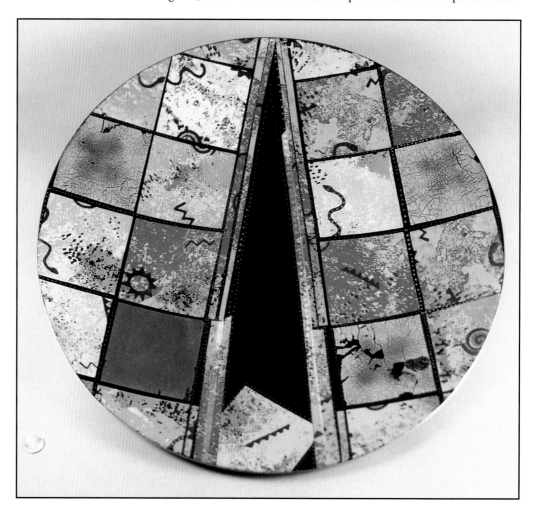

"Shallow Bowl," 18" diameter, enamel on copper, decals, gold and palladium leaf, and lusters.

tion to be sandwiched between layers of enamel. I created and copy-righted two series: "Interchangeable Dress-Up Dolls Necklace Series" and the "Clothing Pins Series." My pieces range in size from 1" to 24".

I first decide what I want to make, perhaps a bowl or a small pin. I usually have my rough sketches all drawn in my head instead of on paper. I begin with a clean, smooth enameled surface. This base coat surface is necessary for the decal to adhere. First, the piece is annealed in the kiln and then pickled to remove firescale, rinsed. Both sides are scrubbed with a grease-removing liquid dish detergent and rinsed. I always enamel the back first with 80 mesh hard black enamel. I use unleaded enamels with a variety of sifters: either 80 mesh and then 60 mesh or 100 mesh and then 80 mesh. Sometimes I paint Scalex, a firescale inhibitor, on the front, let it dry and then sift enamel over Klyr-Fyre on the back. When the piece is fired and cool, the Scalex falls off. I file off any firescale on the edge of the piece. Next, the front is scrubbed and rinsed, and I sift on the enamel.

The decal is soaked in a flat, plastic container filled with warm water to loosen it from its backing. The decal should slip off easily. If not, soak it a little more. My water is relatively clean of harsh chemicals so I do not need to use distilled water. Sometimes I place the decal in the water with the image facing up, which lets me see when the decal lifts from the backing paper, and sometimes I place the decal in the water with the paper side up, to prevent possible excess curling of the decal.

I usually use my fingers instead of tweezers to transfer the decal to the enameled surface. It does not matter if some of the decal hangs over the edge of the piece because you can trim it after it dries. While it is wet, you can move it around, but not excessively. When the decal is in place, the water and air must be removed. Working from the center with a rubber squeegee, a smooth card or a Q-tip, press just hard enough for all the water to move to the edge. Wipe the water away with a tissue or a paper towel. Then place the piece in a dust-free area to dry naturally. If I am in a hurry, I will use either a heat lamp, a warm radiator, or the top of the kiln. When dry, any excess decal is cut away, and the piece is checked for bubbles.

If you see any air bubbles, prick or cut them with a pin or an Exacto knife to release the air and press down the decal in that spot. If the decal is not adhered to the enameled surface, it might not fuse and will not be permanent. The enamel piece, supported on a stilt, is placed in the kiln at

"Charlie Chaplin Tribute" wall piece, 12"h, decals, enamel on copper, fine silver.

"Bridal Couple, Brooch 5," decals, enamel on copper, fine silver.

1450°F to 1500°F to remove the film on the decal, which takes from 1 to 4 minutes. Depending on the size of the piece, there will be a small pop sound inside the kiln. After the combustion takes place, a fine black, gray or brown film comes from the top of the kiln door and may settle on the surface of the piece. This film will burn away. At this point, avoid the temptation to vent or crash the kiln by opening the door and losing temperature.

After the first stage in which the film burns off, colored dyes and oxides from the decal will bond with the liquifying surface of the enamel. Fusion takes place in this stage. After the piece is removed from the kiln, I check for perfect fusion. To do this, I hold the work under a light source and shift the piece from side to side to see whether the surface is even, reflective and glossy. If not, then the piece is re-fired. If you overfire the decals, they will sink into the enamel. This is particularly true of the reds, which always have a lower fusing time. When bonded to the enameled surface, the decal will appear matte. Any transparent enamel or flux can be fired over the fired decal. There seems to be no limit to the number of layers of transparents that can be fired over the decals.

I also use ceramic pencils to create an underlying design element or to put a drawing in place. Ceramic pencils must be applied to a matte surface and then fired. The trick to using pencils is that the surface must have tooth, just like drawing paper has its own rough surface. There are two ways to matte the surface to give it tooth: with an Alundum stone or with matte-salts from a stained glass supply store. Matte-salts come with directions, and the precautions should be followed. After you draw with the ceramic pencils, you can use water to feather out the color, almost like a watercolor. You do not need to cover the surface with enamel, because the matted surface will glaze during the firing process. You must fire these drawings low or you will lose the color.

Sometimes I use lusters as a final coat over the fired enamels. All the ceramic materials that I integrate with my enamels have similar firing temperatures. Some do have a tendency to combust, but everything that one does in the studio with the kiln must be respected or tragedy can befall you. I respect my materials, and they give me fabulous results. I admit that this is only the beginning of what we can learn and apply from the other fire arts. Please continue my search and make beautiful objects to make you proud. ■

"Runners" wall piece, 6"h x 12", detail: decals, enamel on copper, fine silver.

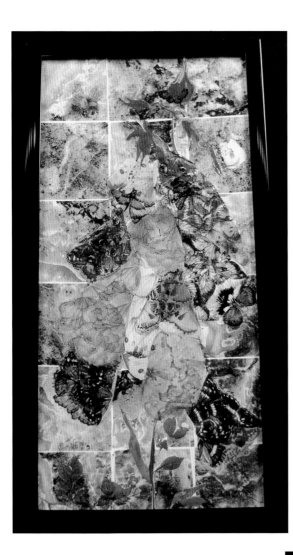

"Counting Flowers on the Wall" wall piece,
12" H x 6" W, enamel on copper, steel,
flowers made with decals in butterfly wings.

"Windows" wall piece,
24"h x 24"w x 2"d,
decals, airbrushed
enamel on copper,
steel.

Enamel Crayons and Watercolors on Steel and Iron

JOHN KILLMASTER

John Killmaster, (b. 1934) Professor Emeritus of Art, has taught at Boise State University since he moved to Idaho in 1970, after receiving his MFA from Cranbrook Academy of Art. Currently he is listed in Who's Who in American Art and Who's Who in the West. Exhibitions of his work include: Denver Art Museum, San Francisco Museum of Modern Art, Smithsonian, American Watercolor Society in New York, and throughout the world.

John Killmaster airbrushing enamel water colors.

"Los Guanajuatenses" wall piece, 9"h x 11.5"w, Limoges enamel crayon drawing, enamel oil, and enamel watercolor washes on steel. *Photo by John Killmaster*

"Sisters" wall piece, 10"h x 8"w, Limoges enamel crayon drawing with enamel oil and enamel watercolor washes on steel. *Photo by John Killmaster*

Many years ago I conceived of the idea for enamel crayons. After very much experimentation, Thompson Enamel Company began to manufacture these crayons, and they were a tremendous breakthrough for the artist. The crayons enable the artist to draw with enamel just like drawing on paper. I use them in combination with airbrushed enamel watercolors.

The majority of my enamel art is on steel or enameling iron because both metals hold their shape with minimal distortion when fired. I primarily use steel because it is available where I live in Idaho. Enameling iron, however, is preferable because it is formulated for enameling. I start with pencil or charcoal sketches on paper and then combine enamel crayons and watercolors with an airbrush on an enameled plaque. If the enamel is an 8" x 10" flat plaque, I use a small sketch to refer to as I draw on the base coated plaque with an enamel crayon. If I wish to duplicate the sketch, I project the image on the plaque and draw directly over it, thus retaining the character of the original drawing. For larger repoussé panels, I can begin by projecting the imagery directly on the base metal and tracing the contours of the image with a permanent marking pen or by cutting out the drawing projected on a sheet of paper. The paper cutout is placed on the metal as a template to be traced around with a permanent marking pen.

I use 16 ga or 18 ga steel for flat tiles, for panels with turned edges and for large repoussé works. The iron is usually 16 ga ordered from a catalog. I purchase 20 ga or 22 ga cold-rolled (low carbon) steel locally at metal supply companies and enameling iron by catalog. The thin metal I use for 6" or smaller tiles and for shaped and hammered pieces. The steel, bought in 4' x 10' sheets, is cut with an electric hand shear or a manually operated metal shear. Shaped works are easily made with shears and jewelry saws. The rough edges are filed before hammering and ground coating. As with all enameling, the metal has to be cleaned first. With steel and iron, a special ground coat needs to be applied and fired before the vitreous enamels.

To prepare the metal for enamel ground coating, I first immerse the piece in Sparex 2 formulated to clean off rust from iron and steel. This is not necessary if the metal is new and coated with oil. However, the oil must be cleaned off with soap or detergent until water does not bead up on the metal surface. Cleanser powder is then applied and the surface is scrubbed with a scrub brush or pad and rinsed with hot tap water. The piece is air dried by standing it vertically against a support.

A ground coat for enameling iron and steel is best applied by an automotive touch-up spray gun. Ground coat #16 Thompson enamel requires thinning with water and straining through an 80 mesh screen to eliminate any large particles. The thinning and screening is essential for all material used with a spray gun or an airbrush. A spray booth with an outside exhaust or a fan and table setup out of doors is necessary for all enamel spraying. Care must be taken not to inhale the spray.

For spraying, I use an air compressor with an adjustable regulator providing up to 25 pounds of pressure. The liquid enamel, used with the spray gun, is diluted about half enamel and half tap water and then strained. I fill the cup half full with the diluted enamel and evenly spray a 1/8" layer of ground coating on the back and let it air dry. With a hot air dryer it takes 10 minutes, whereas in the sun it takes 30 minutes. I then spray the front and let it dry. The piece is then ready to be fired in a pre-heated kiln.

I have a 110V kiln that is 12"w x 12"d for small enamels. The door opens horizontally. For larger pieces I have a home-made 230V kiln with a 24" x 26" chamber. The door of this kiln swings down and is counterbalanced with weights and cables. There are fire bricks on the inside floor of the kiln to support firing screens and trivets. Infinite switches, simple to install, provide manual temperature control on both kilns.

The ground coat is fired at 1500°F to maturity. Additional coats of 250 mesh porcelain enamel (the liquid enamel referred to as base color coats) are sprayed over the fired front ground coat, but not as thickly. When planning to use the airbrush, I prefer the base coat to be white. When using an airbrush combined with enamel crayons and enamel watercolor, I also prefer to work on a white base coat. This coat is fired at 1450°F for about 3 minutes. The length of firing time depends on the size of the piece and the thickness of the metal. Subsequent firings are underfired from 1340°F to 1360°F. I use both the pyrometer and my eye to determine when a piece is ready to be removed from the kiln.

I use the airbrush with the thinned enamel watercolor. Although thinned and screened, porcelain enamel is too coarse for the airbrush. The airbrush I recommend is a dual action, gravity feed with the color cup positioned near the tip above the needle; this one allows easy cleaning and color changing. The airbrush enables an enamelist to produce delicate gradations and extremely subtle shadings of colors and layered overlaps of different colors, approximating art done using watercolor paint. Some practice is necessary to perfect airbrushing.

Enamel watercolor comes in cake form with a water soluble wax binder in it. It is necessary to add drops of water to soften it enough to use with a watercolor brush just like pan watercolor paint. To use enamel watercolor with the airbrush, thin the color with water 1:1, strain it to eliminate any particles and fill the cup half full with color. Begin by adjusting the air pressure to 12 lbs. PSI and pull back the trigger for air while pressing down to release color. Holding the airbrush away from the surface gives a broad spray, while holding it close creates soft linear effects. Almost any tool can be used to remove unwanted sprayed color. If you remove some of the dried enamel with a tool, you will get a sharp line; if you remove the enamel when wet, the line will be blended.

The crayon colors, especially black, brown and yellow, come through the firing process without losing their richness; other colors lose their intensity, but the result can vary widely depending on the manufacturing process and pigments available at the time. The colors available in both crayons and watercolors are black, brown, blue, orange, yellow, green and red. White is available in crayons, and clear is available in watercolors. I would suggest you make some samples of how the colors

"LaMujer" wall piece, 10"h x 8"w, Limoges enamel crayon drawing painted over with enamel oil on steel. *Photo by John Killmaster*

"Jubilee" wall piece, 10"h x 8"w, Limoges, sgraffito, enamel crayon drawing with enamel watercolor washes and oil on steel. *Photo by John Killmaster*

fire. Enamel crayon usually fires to a matte surface. To regain a glossy surface you can cover it with a layer of enamel watercolor in a light color or the clear.

To begin an enamel crayon and airbrush art work, I draw with the black, blue or brown crayon. If you find that the glossy surface resists the crayon, a light layer of hairspray adds some texture that aids the adhesion of the crayon. The hairspray will burn away leaving no trace. You can either fire the drawing or develop the image further before firing by combining several colors, by blending or by dissolving lines with water in a watercolor brush. Enamel watercolors need to be applied twice as thick as the intended effect because firing reduces the enamel and diminishes the intensity of the color by one half. Again, experiments are recommended before you attempt a large piece. Enamel watercolor is fired at 1350°F for about 2-1/2 minutes. **Do not overfire!** If you do, you will lose the colors.

I have found enamel crayons and watercolors very freeing, expressive and applicable to a vast range of possibilities when used alone as well as combined with other types of enamels. I use airbrushes, spray gun, paint brushes, shakers, sgraffito tools, silk screening, alone or in combinations. The possibilities are unlimited! ■

"Triangle Tangle" dimensional wall piece, 12.25"h x 7.225"w x 2"d, Limoges repoussé, enamel oil paint, enamel crayon drawing and watercolor washes on porcelain steel. *Photo by John Killmaster*

"Soul Singer" wall piece, 12"h x 8"w, Limoges enamel palette knife painting over enamel crayon drawing on steel. *Photo by John Killmaster*

RISO SCREEN

JOANN TANZER

Dr. Joann Tanzer, Professor Emeritus of Art, taught at San Diego University for 36 years. In 1960 she developed their enameling degree program. The permanent collections that have her enamels include the Decorative Art Museum, Moscow, Russia; Limoges Museum of Enamels, Limoges, France; Hiroshima City, Japan; and many private collections. Her exhibits include Tokyo, Japan; Limoges, France; Colberg, Germany; Barcelona, Spain, as well as many national juried and invitational exhibitions. She died in 2005.

Traditional photo silk-screen was one of the techniques I taught as a Professor of Art at San Diego State University. The possibility of developing art images seemed endless, but the process of making screens was awkward and time consuming. To develop a library of silk-screen panels in the traditional way required space for storing and time for preparation.

When I attended a demonstration on silk-screening using Riso screen, I was immediately struck with the potential of this process for enameling. The Japanese company that developed the screen used it mainly to make greeting cards. I would like to make it understood that I did not invent this material but rather discovered a use for it for enamels. I found it intriguing because there was an immediacy about it. One could make a screen in seconds. With Riso screen, sometimes called speed screen or thermal screen, it became possible to develop a collection of eight to ten screens in minutes. It did not require an involved preparation of the screen itself, a transparency for the use of the screen nor a bulky wooden frame. Enameling students and artists have been developing additional ways of using this screen.

A screen allows the enamel artist another way of developing images, textures and patterns. Enamel falls onto prepared surfaces through the screen to make a precise image. The use of the screen can be ap-

"A Code of its Own," 10" x 10", enamel on steel, silkscreened.

plied to any style or attitude of art. I do not often use a screen as a single image, but as a series of small segments together to enrich the composition. The potential of this technique is unlimited. It can have an open, loose approach as well as a precise, rigid line and it can be changed in a myriad of ways.

I am fond of using collage images because they are so serendipitous. I cut out quantities of material from newspapers, magazine and even throwaway cartons such as cereal boxes. When I view my collection and make aesthetic judgments for a particular image, I arrange the clippings on a stiff paper surface. You also can use any technique, for example, pencil, ink, photos, to make the original image independently or in combination for a variety of patterns, markings and textures, as long as the final art work is copied on a graphite copier. These other techniques also can be combined with the clippings for a collage.

Once I have chosen the image, I begin the screenmaking process by Xeroxing the collage to make a graphite image that will become the pattern of the actual screen. Any copier that uses graphite can be used. The image to be copied cannot have large areas of black because the black on the photocopy will stick to the screen. If you want to use a large black image, run the copy through the copy machine several times in order to lighten the area and have a satisfactory screen.

What is Riso screen? It is a green nylon mesh material that is both a screen and a film that is heat sensitive. It comes in rolls of various widths and several different meshes and also by the sheet. The roll comes with a plastic sheet in the package to support the photocopy that is now your art work along with the screen to go through a thermal copier. By the sheet, the green screen comes with a piece of white paper attached at both short ends.

You make the sandwich to go through the thermal copy machine with everything facing up. On top is the rough side of the green mesh, the smooth underside etches your carbon photocopy into the screen; under the mesh sheet is your photocopy, right side up against the smooth side of the screen; and under the photocopy is either the plastic sheet or a piece of white paper as a support. The Riso screen sheet, about 9" x 11 ½" and looks like a sheet of shiny green paper that is attached at the short ends to a piece of white paper; the two sheets are open on the two long sides.

I usually buy the Riso screen in 70 or 100 mesh rolls. Welsh Products will sell single sheets of Riso. I advise beginners to start with the 70 mesh, which allows the 80 mesh enamel to pass through more easily than the 100 mesh screen does. The finer mesh gives more precise detail. The dimensions of the screen you use depend on the opening in your thermal copier and the size of your kiln. A new thermal copier on the market is Vistafax. You may be able to find a second hand copy machine. I have several machines. They range from a portable unit to one with a 24" opening. I even found one for $5 in a junkyard.

With a small piece of screen make a test run with an image that fits the scrap. The sandwich is run through the thermal copier with the green side, the shiny side, up. The proper setting on the 3M Thermofax varies with each machine. Start with #5 on the dial; move the setting up or down until you get a clear image. Feed your blank paper, image and screen into the thermal machine at the proper setting. The sandwich will return to you automatically. I usually run it through two or three times to check for clarity and a strong, crisp image. I check the screen by folding back one corner and holding it up to the light. If the screen is cut properly, you will see a sharp imprint of your image.

The screen may be used on bare metal that accepts enamel or on a flat piece of pre-enameled copper, fine silver or steel. Before you work on a large piece, I suggest you make some samples on either 2" square or 4" square tiles. I assume you know about various metals, enamels and firing. For base coats, I use medium or hard, light color opaques or hard flux.

The Riso screen needs to be in a frame that will hold it taut for the enamel to be worked through it. The frame can be purchased in plastic or cardboard, which is not desirable for wet enamels. The frame will hold the screen about 1/16" above the metal or enameled tile so as not to disturb the enamel when the screen is removed. When I make a frame, I usually attach a lifter to the frame to help. The lifter is a spacer made of bits of very thin cardboard or other material. I like to make my frame with a 14" embroidery hoop. I also purchase a Welsh Products plastic adjustable frame, which I use raised up 1/16" for dry screening.

To use an embroidery hoop, take the finished screen and center it on an 18" square piece of cotton fabric or rip cloth, which dries very fast if it gets wet. Leaving enough material to stretch the cloth taut into the hoop, sew the material onto the screen around the edge. Then cut out the section of material under the screen.

For dry screening, I use 80 mesh, unwashed medium fusing enamels and sometimes soft enamels for the final coat. I put the screen in a Welsh Products plastic frame that lifts it 1/16" above the metal. You can use either a stiff 3" x 5" card, a plastic credit card or a commercial squeegee to push the enamel through the screen onto the flat metal or enameled surface. Carefully lift off the screen and place the piece onto stilts or whatever you use to support the piece in the kiln. Fire at 1500⁰F until the enamel is fused, remove the piece from the kiln and weight it if necessary.

For wet enamel, I prefer using the screen sewed to fabric and stretched in the embroidery hoop. The plastic frame is satisfactory also. I use dry crackle color that I mix with water to the consistency of cold cream. I push the enamel through the screen with a rubber or plastic cot on my forefinger; you could use a tool. The wet enamel gives you great control, and the colors can be quickly changed or mixed on the screen. If you wash the screen very soon after using, it will stay in better condition. The washing will not injure the screen, but do it gently in order not to damage the image. Many layers of the wet surface can be applied before firing, but allow a drying period between coats of the screened image. It is, however, advisable to fire between layers of screening with wet enamel; allow the enamel to dry completely before firing. The screen may be allowed to dry, dried in the sun or with a hair dryer. A great deal of editing can be done after the enamel has dried on the panel. You can sgraffito the enamel or remove areas with a brush, stick or card.

Here are some suggestions on ways to use the screen on 2" square copper samples:

1.   Apply opaque enamel >> fire >> screen opaque image in contrasting color.

2.   Apply light opaque enamel >> fire >> screen dark transparent color.

3.   Apply flux >> fire >> screen opaque dark color.

4.   Apply light transparent color >> fire >> screen dark transparent color.

5.   Screen transparent image on raw copper >> fire >> apply a light transparent >> fire.

6.   Then try all of the above in multiples of overlays. Using a single screen image, juxtapose the image over and over, still retaining the feeling of the original. Explore with color themes and relationships to achieve desired results.

Screens do not need to be stored in their fames but do need a cool to moderate temperature. You will be able to save and keep many screens in a small space, such as a loose-leaf binder with plastic sleeves for the screens. The screens can be folded for storage to be used over and over. As you accumulate a collection of screens, you will find their versatility endless. The possibilities are only limited by your imagination and your knowledge of how the various enamels fire. ■

OTHER TECHNIQUES

Past Commander Pin, three sections, 4.75" x 1.5", 14k and 10k gold: engraved, die struck, cast, and fabricated; enamel, champlevé.

TORCH FIRING ON GOLD ALLOYS: CHAMPLEVÉ AND BASSE-TAILLE

EDWARD J. FRIEDMAN

Edward Friedman, master platinum smith, has been designing and fabricating jewelry for over 25 years. He taught for Rio Grande and Precious Metals West. He is a Senior Instructor of jewelry arts at The Revere Academy of Jewelry Arts, San Francisco, CA. where he also teaches enameling. He owns The Buehn Company, a Masonic manufacturing company, which introduced him to enameling in 1997. He is also master model maker and manager, Casting Division, Hoover & Strong, Richmond, VA.

Champlevé and basse-taille are two enameling techniques that are often used together. In basse-taille, the metal has a design that can be made in various ways. The textured metal is covered with transparents to allow the design to be seen. I make a design in the metal with gravers or texturing punches. The champlevé technique ordinarily uses opaque enamel in the cell. However, if the bottom of these cells is textured, it is enameled with transparents. I use both of these enameling techniques in the manufacturing of rings, pendants, jewels, wedding rings and bands. My work is done on 10k, 14k and 18k in both yellow and white gold as well as platinum.

I purchase standard sheet and casting grains containing about 2% zinc. In trade work, many clients send cast pieces to be enameled. When I do not know the alloy of the metal, it is very important to clean the metal and depletion gild it after the soldering and before enameling. I either fabricate or cast my piece to be enameled and then anneal it to soften it for engraving. Next I lay out the cells, which should not be more than .6 mm deep. If the cells are deeper, the enamel will crack due to the different rates of cooling between the metal and the enamel. Many pieces are die struck to create the cell areas.

For hand engraving the outline of a cell, I use a sharp onglette graver about .3 mm deep. Next, using a flat graver the same size as the width of the gap between the onglette cuts, I lower the base of the cell to the desired depth. It is important to keep sharp crisp corners and a flat bottom in the cell. To texture the bottom of the cell, I use either the same flat graver to do a wiggle cut or I use a Florentine graver.

It is important that all soldering is completed before enameling. Some pieces, such as catches for pin backs, bails or jump rings need to be soldered with the highest melting temperature solder for that alloy. Many of my pieces have at least five separate enameled sections that will need to be assembled to complete the finished jewel. These enameled pieces need to be tube set from the back. The tubes must be soldered in place prior to enameling. After all soldering is completed, all pieces must be depletion gilded, which removes any copper from the surface of the metal and leaves a layer of fine gold.

To depletion gild, you must anneal the metal without a fire coat by first coating the piece with boric acid and alcohol solution, torching the piece, letting it cool until all red color leaves the metal, and then quenching the piece in water. The piece is pickled in Sparex2 for several minutes until all oxides are removed and then it is rinsed in water. I use a fine brass brush to clean the piece before placing it into a cool nitric acid bath for no more than one minute. It is rinsed and placed in the ultrasonic for further cleaning. The entire depletion process is repeated three to four times to remove all copper from the surface of the metal. The process is complete when the surface has an appearance of a muddy high karat film over the entire piece. If copper is left on the surface, it can come into contact with the enamel and change color. White, red and green are the most susceptible to this effect in both opaque and transparent enamels.

I purchase leaded enamels in 80 mesh. The transparents in leaded enamels have a deeper color and tend not to crack as often as unleaded enamels. Powdered enamels should be kept in airtight opaque containers to protect them from air and light. I prepare only the amount of enamel needed for the day's work. First I grind the enamel with a mortar and pestle to a fine grainy consistency. Then I wash it with distilled water until the water is clear. This result takes about eight washings. The washed enamel is placed in a small container and is wetted with distilled water when it is used.

To wet-lay the enamel, I use a scoop that I made from an old file. I hammered one end into a spoon shape about 1 mm in diameter. I do not use gums, but lick the object because the saliva acts as an adhesive. I do not counter enamel the champlevé pieces because the layer of enamel in the cells is relatively thin.

The first layer of enamel fills the cell to excess to ensure a full fill and to protect the surface of the metal. As the enamel flows over the surface, it covers the entire area around the cell. The enamel keeps the oxides from forming in the metal area in and around the cell. Using the scoop, I tap the edge of the piece to pack the enamel grains tightly together. The piece is set aside to air dry before firing.

About half of my enameling is with the torch, and the other half is kiln enameled. I use an oxygen/propane mix with a Meco torch with a #4 tip for most of my torch enameling. I hold the piece in the air with an insulated spring tweezers while heating from the under side to bring the enamel to its flowing point. I have 2 kilns: one is 3" x 3" x 4" and the other is 4-1/2" x 4-1/2" x 6" with a pyrometer. I fire at 1500°F to fuse the enamel to the metal quickly. So not to melt the metal, I set a timer to one minute, which gives me about five seconds before a complete flow of the enamel occurs. I fire all firings to maturity.

After each firing, the piece is placed on a warm charcoal block with a watchmaker's tin on top to cool. For a larger piece, I use an Altoids (small candy) tin with the lid removed. This step ensures that the piece cools slowly and so minimizes the chance of the enamel cracking. During the cooling process, some of the enamel may chip off and need to be refilled. Refilling will have to be done two to five times to bring the enamel flush with the surface of the metal. Between each repacking, the enamel is ground down smooth and level with the surface of the metal, and the piece is placed in the ultrasonic cleaner and the steamer. This cleaning process removes any particles of ground enamel and any parts of the grinding stone from the crevices and cracks that could contaminate the next firing. I use a diamond hone of medium grade.

When the cell is fully enameled, the piece is placed in a Sparex2 pickle solution for no more than a couple of minutes to remove the oxides. If you leave the piece in the pickle too long, the acid will etch the enamel and, in turn, the enamel will have to be removed. Removing the enamel means that the whole process will have to start again from the beginning. Next comes sanding the metal and enamel with a 400 Corundum and a 4/0 emery sandpaper. The sanding gives the surface of the metal a semi-polished finish and makes the final polishing easier. The process of stoning scratches the metal, so the metal must be smoothed before the final firing.

After the pickle, I rinse the piece and finish cleaning with a glass brush, which will not scratch the surface of the enamel. Prior to any enameling between firings, always use the ultrasonic and steamer to remove all particles in the cell and on the enamels, ensuring a clean surface prior to the next fusing. If you do not have a commercial steamer, you can use a cappuccino maker for steam.

The enameled surface should be smooth after sanding with the 4/0 emery paper. Occasionally the surface of the enamel may be rough and uneven. If so, I use a 6" round, 1" wide lap or hard felt wheel with a catch tray underneath. The catch tray is filled with pumice flower powder mixed with water, forming a paste that should have no lumps. The lap is charged to polish the surface of the metal and the enamel.

After the numerous firings, the metal will have a thick oxide scale that must be removed. Pickling does not remove this scale. To remove it, I use bobbing compound on a 3/4" boar bristle brush in my flexshaft. To finish the outside of the piece, I use a 6" yellow muslin buff on my polishing lathe. Rinse and finish the piece with red rouge, polishing to a final luster.■

LARGE MOSAIC ENAMELS

JEAN FOSTER JENKINS

Jean Jenkins was professor of jewelry, design and enameling at El Camino College, Torrance, CA. She received a BFA in Art and Education, 1943 Carnegie-Mellon; MA in Art, UCLA 1967; Cranbrook 1975. Her first enameling instructor was Kenneth Bates, Cleveland Art Institute. Jean's focus has been on mosaic enamel wall pieces ranging up to 5' x 7'. Her enamel work has been shown nationally and internationally. Her studio is in Palo Alto, California.

Jean F. Jenkins

An enamel of any size can be produced by working in sections. I cut copper sheet into irregular shapes to enhance and strengthen a design and to give the illusion of a large single piece. The pieces are mounted on a single sheet of 3/4" marine plywood. This method requires a worktable large enough to lay out the full-scale work. I have made many panels by this method, the largest being 5' x 7', which is the size of my worktable. The large panels are hung with the two-piece brackets used for wall cabinets.

I first design a maquette about one-third the size of the planned enamel mosaic. It is an accurate full-color depiction, to scale. Then I make a cartoon (a line drawing) to plan the cutting of the sections of the mosaic and number each piece. The cartoon should be based on the abstract design of your composition and divided into smaller pieces as necessary. Avoid long, thin shapes, C shapes or S shapes and highly irregular shapes. With a pantograph, I enlarge the cartoon to full size on wide butcher paper. I then make any necessary corrections and number the pieces to match the first cartoon. Each piece should be able to fit comfortably in your kiln with a 2" margin on all sides.

"Allegory—Silicon Valley" wall piece, 5' x 7', mosaic, sifting and wet-packing enamel on 18 ga copper pieces. *Photo by artist*

"Self Portrait" wall piece, 4.5" x 6.5", drawing, sifting, wet-packing on 18 ga copper. *Photo by artist*

I first work on the plywood sheet to which the enameled sections will be attached. Wood sealer is applied to the back and edges of the plywood sheet. When dry, square (3/4" to 1") aluminum tubing is cut to form an X shape to be screwed to the back to ensure the rigidity of the panel. Allow clearance for the frame. These aluminum tubing pieces should be drilled to accept wood screws that will go through the tubing and halfway through the plywood. The X shape is positioned on the plywood and screwed in place. The plywood is turned over to accept the tracing of the cartoon and then the copper pieces. The full-scale cartoon, including the numbers, is traced onto the plywood with a fabric marker wheel and carbon paper. It is placed on the plywood and fastened in place on two sides with drafting tape.

I buy large pieces of 18 ga industrial scrap copper, cold-rolled, annealed if possible. With a Beverly shear, an electric shear or even a jeweler's saw, I cut the copper after I have cleaned it to remove any oil or dirt. The cartoon pieces are glued with rubber cement one at a time to a piece of copper. After the copper is cut to shape, its number is put on the back with an electric engraver. The rubber cement is removed from the back of the cartoon piece, and the cartoon piece is saved in a folder for reference. The copper pieces are flattened with a rubber or leather mallet on an anvil.

I clean the metal in a pickle of the standard Sparex2 solution to remove the copper oxide, rinse well and scour with steel wood and detergent. When the water sheets off the surface, the metal is grease-free. I wear latex or plastic gloves to protect my hands and to protect the metal from fingerprints.

I use primarily leaded enamels from my inventory in 80 mesh and 150 mesh. I use 150 mesh when the piece requires four or more coats. I apply the enamels by either sifting or wet packing and use various techniques depending on my design.

"3 Kings" wall piece, 2.5' x 3.5', mosaic, sifting, wet-packing, enamel on 18 ga copper. *Photo by artist*

For the base coats, I use liquid hard flux with backing enamel for counter and 80 mesh medium flux sifted over a solution of metho gum on the front. The gum I use is carbo-metho-cellulose (CMC) from Hercules Powder Co., Wilmington, Delaware; 52 grams to make 5 gallons. To make it into solution, I dissolve all 52 grams in 1 gallon of distilled water and then mix in another 4 gallons of distilled water.

Smooth, even coats of counter enamel on the back will discourage warping. The irregular shapes and larger pieces are fired on the "bed-of-nails" stilt. My kilns have a pyrometer and a rheostat. The kiln is preheated to 1550°F before placing the piece in the center of the kiln with small areas or angles nearest to the door, away from the elements. I check the firing by opening the door a crack. After the piece is fired, it is weighted with a press plate on a warm marble slab to keep it flat. The hard fusing enamels are fired first, the opaque reds and oranges added close to the last firing.

After every firing, the loose scale from the edge of the copper piece is removed with a medium-cut metal file. Each piece is placed on the cartoon as I work. For efficiency, I enamel all pieces of the same color first. They are sifted, fired and, when cool, are placed where they belong on the plywood. I plan all the colors initially. Wherever I plan to use silver foil, I sift and fire 80 mesh medium fusing white on that area. The foil is cut and placed on a sheet of static preventative before being positioned with diluted gum; it is fired when dry.

The biggest problem with enameling copper is that the copper grows with each firing. The copper expands in the heat, and the glass solidifies before the copper has returned to its original size. Therefore, I use my cartoon pattern to check the size of each piece. If any edge needs to be ground off, I trace that edge with a permanent marker. I grind as needed and keep each piece its proper size.

My motor has an exhaust hood for an expandable rubber wheel with a coarse (80 grit) wet-or-dry abrasive belt. The wheel is kept wet with a spray bottle of water to prevent heating and dust. I wear a particle mask. I do not grind or stone the enamels for a smooth finish: I feel the irregularities of enamel thickness enhance the visual impact.

When all the pieces have been well developed, I climb a tall step-ladder to look down on the work as a whole to see where color value changes are needed. The design grows, as an easel painting would develop. I keep complete notes on each piece of the cartoon pattern.

For indoor installation, I use hot glue to attach the enamel pieces to the plywood. Next, I apply non-sanded grout, formulated for kitchens and bathrooms. Oxides are added to the grout for smooth blending from one section to another. The grout is applied with a small artist's spatula; I clean any excess as I go. As a fihnal step, the grout is covered with grout sealer. The mosaic is then ready for reaming and/or installation. The pieces must be set and grouted with mortar if the installation is to be outdoors.

I have focused on the way I work today. Tomorrow may bring a different story, so don't be afraid to be creative. ■

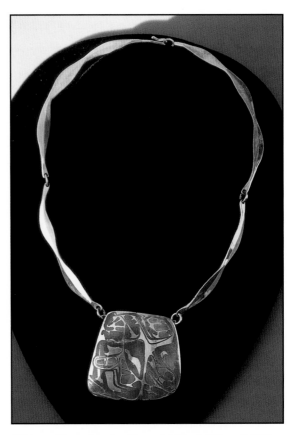

"Tidepool" pendant/necklace, 1.5" to 2.25"w x 2.25"h, champlevé wet-packed enamel on etched fine silver, necklace: hand forged sterling silver. *Photo by artist*

"Chartreuse" necklace, enamel pieces 1.25" x 1.5", basse-taille enamel on etched fine silver, setting: gold plated, 10 pierced sterling silver pieces complete the necklace. *Photo by artist*

"Celtic Sunrise" wall piece, 3' x 6', mosaic, sifting, wet-packing, enamel on 18 ga copper. Jean Jenkins in photo.

TORCH FIRING

DEBORAH LOZIER

Deborah Lozier is a full-time metalsmith in Oakland, California. She received her BFA from Arizona State University, Tempe in 1984. She has focused on developing enamel and patina torch firing on copper forms. She teaches and exhibits internationally. Her work is in the permanent collection of the Oakland Museum of California. Publications include Ornament, Metalsmith, American Craft *and the book* Color on Metal.

Deborah Lozier in studio. *Photo by Eric Smith*

Torch firing is a wonderful way to experience enameling in all of it stages. I see it as an extension to kiln fired enamels and not a replacement since the results are very different. It requires working from instinct with a sensitive observation of cause and effect. The torch oxidizes and blends the pigments, creates patterns and causes chemical reactions that do not occur with kiln firing. Firing times are intuitive, and the temperature is controlled by the length of the flame. The torch fires with a cascade of heat, allowing for control over heat placement throughout the piece. Solder seams and delicate colors (soft enamels) thus can be avoided while areas needing high heat can be hit directly.

This flexibility requires active participation to not overheat delicate areas and to adequately heat the harder enamels. The torch, creating an active heat, causes the enamel to move while the fuel oxidizes the pigments and in turn mixes with the oxides contained in the metal. In thin applications, transparents will shimmer and opaques will take on varying degrees of translucency with glaze-like qualities. In thicker applications, they will become stronger and appear like stone inlay. The more direct contact with the flame, the more exaggerated these effects become.

Heating the piece from underneath creates effects closer to kiln firing, but also limits size potentials. The workable size depends on the heat available from the torch instead of the perimeter of the kiln. I always experiment with the direct flame in mind since color variables differ greatly. Torch firing does have its limits, but also offers possibilities that give a spontaneous and open approach to a historically reserved discipline.

"Wedding Crown" sculpture, 3.75" x 4" x 4" Limoges torch-fired enamel on copper. *Photo by artist*

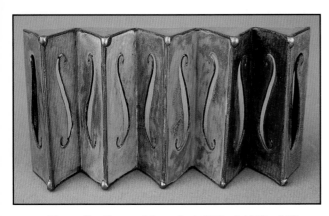

"Accordion/S-curve" brooch, 1.625" x 3.125" x 0.5", Limoges torch-fired enamel on copper. *Photo by artist*

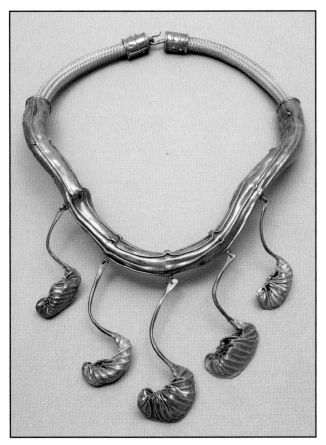

"Suspended In Time" necklace, 11" x 8" x 1", sifted, torch-fired enamel and patina on copper, brass. *Photo by Ken Rice*

TOOLS AND MATERIALS

Eyewear is the most important element for safety. I use Auralens glasses, made for glass blowers, and can be ordered with a prescription. They cost around $180, but provide protection from ultraviolet light, infrared light and sodium flare while allowing good visibility. I use them even for soldering. I also wear leather gloves, an apron and close-fitting natural fiber clothing (no polyester, no sandals). My firing station is equipped with a roof-mounted ventilator to remove fumes. No flammable items, such as papers or cloths, are near the firing area. A good beginning torch set-up is a Smith or Presto-lite acetylene atmosphere regulated torch. Get the B tank not the little mc. The #2 torch will be used the most, but buy as many different tip sizes as you can afford. Oxygen/acetylene is too hot. The little butane hand torches are not hot enough, but if you own them, try them. My top torch recommendation is natural gas and oxygen if you are experienced at using torches. I have two torches with a variety of tips: the National Hand Torch and the Unecon Hand Torch. The Unecon has an automatic shut-off lever. The tips are interchangeable. They can both be used with propane and oxygen. They were bought from Wale Apparatus Company, a glass-working supplier. Smith also makes a nice set-up.

The basic #1 to #5 silversmithing tips are good for jewelry scale pieces. For larger scale pieces I use Wale's hush tip that comes in three sizes. These tips create a bushy rather than a focused flame and they are good for heating a general area. Smith offers a tip, called a high heat tip, which should offer similar results.

Many of my firings require two torches. I use an automatic striker to light the torch with one hand. When two torches are being used, the torch hook and automatic shut-off lever become important. For safe two torch firings, I have bench hooks to hold the torches. Having two large torches going with the ability to turn one of them off quickly is safer. Obviously, limit yourself to one torch until you are experienced before trying to use two at once. Until then, use an assistant to handle the second torch.

The firing station is fairly simple. I use a tripod, a ring stand, a rotating annealing pan, three-point-open-bottom trivets, and open weave stainless steel screens. The screen of 18 ga to 20 ga wire and woven at least 1/4" apart is available from many suppliers. I modify soup cans to create trivet stands and use an annealing pan so I can rotate the piece during firing. I use large steel bolts and washers to raise up the tiny trivets when I am firing rings. Steel rods can be bent into holders and stands for odd shaped pieces. The stainless steel screen, on a tripod or ring stand, works for smaller pieces that are not enameled on the back, such as an earring or pin. I attach 18 to 20 ga steel binding wire to create a simple, non-stick lift for small pieces enameled on all sides. Very small pieces can be held directly with a pair of holding tweezers if the pieces have a secure appendage to grasp. I also use props sold for cooking over campfires and home barbecues along with a variety of household items that I alter to meet the need. The firing solutions are an important part of the designing process and require constant modifications.

I use all Thompson's unleaded enamels except for their five C-grade transparents, which cast a gray blue haze across the surface unless they are only fired from underneath. Leaded enamels are unsuitable because the lead tends to rise to the surface as a gray haze. I originally used just the 80 mesh, but I now screen the 80 mesh enamels to have the 325 mesh for painting and intricate champlevé and the 200 mesh for a more even sifting. Many of the liquid enamels are appropriate for torch firing but respond best to a natural gas and oxygen torch set-up. Their fine grains and thin applications make them susceptible to oxidation and burning away.

Any metal that can be enameled in a kiln can also be torch fired. I prefer working with copper or fine silver. Copper has a lively color personality of its own and a high melting point, which allows flexibility for heat placement. I do not wash the enamels, but I do try to keep the work area clean. The application tools and procedures are basically the same as kiln firing. I clean the base metal, use enamel adhesive for sifting and wet techniques, allow the piece to dry and then apply heat. All the techniques can be incorporated, realizing that torch firing creates different color qualities. Beautiful abstract surfaces can be achieved. The uneven heat and contaminates in the fuel will not produce the pure color and density found in contemporary cloisonné work. Plique-à-jour is not an option as there is not enough metal surface area. With the torch, the enamel and oxides move with the heat, which causes lines to blend and colors to flow.

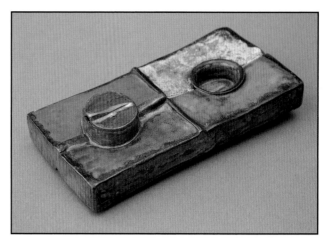

"Circles & Rectangles" brooch, 2.75" x 1.75" x 0.5", Limoges torch-fired enamel and patina on copper. *Photo by artist*

FIRING AND DESIGN

The main difference in designing for torch firing lies in creating areas for direct torch access. In its powdered state, the enamel will burn with direct contact with the flame. For color clarity and proper fusing, the enamel initially requires indirect heat from the metal. This means there must always be an available area of bare metal or previously fired enamel to start the torch firing process. If the entire surface of a piece is to be enameled, it must be treated as if it has a front and a back or an inside and an outside, then fired in succession, always leaving a place for the initial torch heat. For example, if I am firing a bowl, I first apply enamel to the inside of the bowl and direct the flame to the bare metal on the outside. For the next firing, I apply enamel to the outside and direct the flame onto the fired enamel on the inside. I continue to rotate the applications and firings until the piece is complete. I usually do this loosely, with two or three firings in a row on the inside, then moving on to the outside for a few firings. I rinse the piece in water between firings and do not pickle it unless there is an unenameled area from a previous firing that I want free of oxides for a transparent layer.

I plan applying the hard enamels first and in areas that will receive a lot of direct heat while the softer enamels are built up slowly or placed in sheltered areas of the design. Most colors will change with each heating. Many colors improve with increased firings. It takes experience to learn which ones. Color quality will be best if pieces are fired rather quickly. This requires an active and mindful maximize-the-heat technique somewhat similar to soldering. The fusing will follow the heat. The heat will be greatest any time the torch can be placed perpendicular to the surface. I use a bushy flame as for annealing. After the enamel is applied and dry, I begin fusing by heating either a bare metal area or a previously fired enamel layer. With a relaxed spiral rhythm, I gather the heat in an area to start the fusing process. Once an area begins to respond, I focus the heat on that area and then expand it by spiraling the flame throughout the piece. On something small, the response will happen quickly and easily. I can choose to finish the fusion from under neath, or decide to transfer the flame onto the enamel surface to finish the firing.

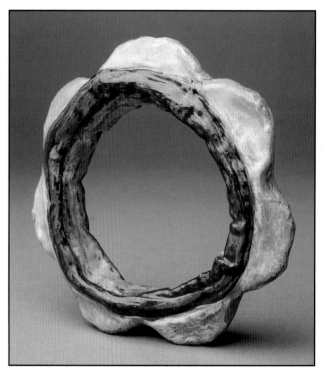

"Folded Flower" bracelet, 4" x 4" x 0.625", Limoges torch-fired enamel on copper. *Photo by artist*

Large hollow forms will require: two torches, a more acute observation of the fusing taking place and flame contact over the entire surface of the work. Once the fusion begins, I move one of the torches to follow the fusion as it spreads and use the other torch to retain an overall constant temperature. At any scale, a consistent temperature is desired—not too hot, not too cool. Be your own thermostat by moving the torch flame away from the piece periodically so that overheating does not occur.

Heating is also dependent on other variables, including the shape of the piece, the gauge of the metal and the type of trivet or stand it is resting on. A 24-ga pair of earrings, resting on a 16 ga wire screen, will fire easily, using a small to medium soldering flame, in a minute or two. In contrast, an 18-ga bowl resting on a trivet may take more than five minutes with two torches going.

Complete heating is a problem with hollow forms, especially those with a large enclosed volume, and causes some size restrictions. I use a tempo-

"Curved Line As Negative Space" sculpture, 7" x 5.5" x 2", Limoges, sifted, torch-fired enamel and patina on copper. *Photo by artist*

Firing area: small to medium scale torch firing set-up with props. *Photo by artist*

rary covering for the initial phase of firing to help contain the heat. A tin can with a small window cut out for viewing, resolves the problem for small hollow forms, and a structure can easily by fabricated out of ductwork parts or stainless steel sheet for larger work. Once the enamel begins to fuse, I remove the cover with tongs for complete torch access. When two torches are being used, the torch hook and automatic shut-off lever become important. The tools and firing set-up must be carefully maintained and respected for safe operations. Safety should always come first. The number of firings is piece specific and subjective. More firings usually allow for color depth and intrigue. On an average, I fire pieces 8 to 12 times to orchestrate harmony in the color ranges. As I said before, you need to experiment. There are a few opaques and transparents that are not compatible due to the differences in expansion and contraction. The different fuels create color differences that are more pronounced when the flame has direct contact with the enamel. Transparents will shimmer and clear with intense heat; opaques oxidize more noticeably and many require more thoughtful planning. A clean yellow or orange can be hard to achieve, but if you apply these colors in areas protected from the direct flame, adding oxides to pump up the pigment content, or apply them at the end of the firing sequences it will help. Transparents layered over opaques can protect the opaques from adverse effects and usually improve the intensity and depth of all colors.

The thickness of the layers also affects the color quality. The Thompson #533 liquid white, when hit with a direct flame, creates an incredible rusty orange and beige surface, while this same heat on a thin layer of green or yellow turns dark and unappealing. As the layers are built up, this effect lessens. The copper can also have a first layer of flux to keep the metal oxides from interacting and darkening the color. Keep in mind that transparents tend to become more brilliant with increased firings, even if no fresh enamel is applied, and opaques will weather and fade. You need to plan ahead once you have learned the traits and needs of the individual enamel colors so that colors needing high heat can receive it and those needing a delicate touch can be in protected areas or receive a final light heating. I keep notes of my observations and find them an essential tool.

I solder pieces to construct an object and then enamel it. Torch firing the enamel, when done carefully, can protect the seams from opening. I usually use hard solder and occasionally medium solder. Since solder will flow when heated to its specific temperature, it is necessary to pay attention to the seams and resist applying too much heat to them. If I choose to apply enamel next to a soldered seam, I direct the heat to the seam last and gently pull the torch back to a cooler point of the flame and stop firing just as the enamel is fused, but not a second longer. A lap seam will fare better than a butt seam because it provides more surface area for the solder and allows a little movement without failing. Edges can be rolled over seams or rivets can be added to strengthen them. Many forms can be easily constructed using box making techniques or die forming, all creating very durable connections.

I design appendages with extra surface area for the solder or design them into the body of the piece. I never rest the weight of a piece on an added element during the firing because that would risk pulling the seam apart. I rarely solder after enameling. I have now begun to weld seams for large pieces rather than solder them. The welded seam can be covered with enamel. The seam is not as crisp. Tack welding can also be used with soldering to secure a long seam and maintain a crisp joint line.

FINAL THOUGHTS

I listen to the material's natural potential and set up the work to encourage these natural tendencies to happen and with more predictability. There is control, but I leave a wide margin for error and do allow some of the errors to remain. An accident that can be recreated turns into a technique. Space between the action and the reaction is where discoveries are made, and I work to be ready for their arrival.■

LAYERING OVER SGRAFFITOED LIQUID ENAMEL BASE COAT

JUDY STONE

Judy Stone lives in the San Francisco Bay Area and has been enameling professionally since 1972. She supports herself from the sale of her work. Her life revolves around enameling. She teaches it. She advocates for it. She probably dreams about it, although she never can remember her dreams. The imagery in her work, while mostly abstract, has reference to floating forms lurking in her subconscious. She has developed her own way of working based on availability of enameling supplies and the contemporary work of the late Fred Ball.

Judy Stone. *Photo by Phyllis Christopher*

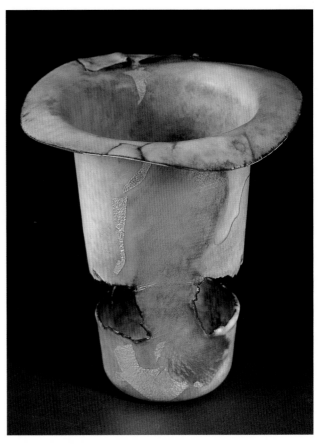

"Burnt Offering VIII," 18 ga copper, spun vessel form, plasma cut, riveted, gold foil, satin finish enamel.

I make light switch covers that are site-specific functional fine art and sculptural vessels that are not at all functional. I sgraffito my design into dried Thompson's #533 liquid white enamel base coat. I layer transparent, opaque and opalescent enamels over it in many firings. The white liquid enamel base coat over the metal is similar to a coat of gesso over canvas in oil painting. In subsequent firings, the layers go from transparent to opaque or dark to light.

I usually enamel on 18-ga copper. I form my switch cover shapes from copper sheet pressed in a hydraulic press with a Masonite die. I either buy copper vessel shapes and alter them or I fabricate them from copper sheets. I like either to cut holes in the vessels or cut pieces out of them and then sew the pieces with copper wire or rivet copper pieces over the holes.

The copper is cleaned by firing the pieces in the kiln for about a minute at 1500°F. After the copper cools, I remove any loose firescale with a toothbrush. The base coats, front and back, are applied with an airbrush in the spray booth. I usually clean and spray several pieces at the same time. Thompson's #533 liquid white needs to be stirred until all lumps are dissolved and also diluted with water, if need be, to the right consistency for the airbrush. I prefer the even coating the airbrush can achieve. An uneven application usually results in a myriad of colors and textures. I spray only a small amount of the liquid enamel at a time because the heavier enamel particles tend to sink and must be agitated to keep them dispersed in solution. The airbrush needle is set way back in its tip, and the aperture is wide open to keep the heavy enamel material flowing evenly through the airbrush. When all pieces are coated, they are dried with a heat gun. Industrially, this is called "bisquing" because drying at 1200°F creates a very hard surface.

The hard surface of dried, white, liquid enamel is sgraffitoed with a sharpened bamboo skewer. The pieces become drawings, sometimes with very noticeable strokes, sometimes with different width lines or with large areas of exposed copper created by first scratching with the skewer and then brushing the enamel away with a soft brush. If I am working on the three-dimensional vessel forms I will spray all sides with #533 and sgraffito all sides. If I am working on switch covers, I will brush a Thompson liquid counter enamel on the backs of the covers after sgraffitoing the fronts and then sift 80 and 100 mesh enamel over the still wet counter enamel.

"Burnt Offering XV1" vessel, 6.5" diameter x 3"deep, Limoges sgraffito enamel on spun, cut, riveted, and sewn copper form, fine silver foil, satin finish. *Photo by Ralph Gabriner*

The switch covers, with both sides coated for the first firing, are fired with the white side down on a point rack. The points are coated with kiln wash and allowed to dry before they are used. The vessels are fired concave side down on either a trivet or a fired support constructed for the piece. I fire to maturity at around 1500°F. The kiln I use is an old 220V Vcella with a fairly large chamber. It is very energy efficient and retains an even heat.

After the first firing, the pieces are put in a Sparex2 bath to remove the firescale that has formed over the exposed copper lines and spaces, then rinsed well and dried. My procedure of layering has developed as my knowledge of enameling has grown and as I have learned about the three-dimensional aspects of the color I am working with. The first layer of enamel over my sgraffitoed surface is a sifting with a 150 mesh sifter of never more than four or five colors of Thompson's 150 mesh lead free transparents. I overlap some of the transparent enamels to suggest an under-painting that I will develop with subsequent layers. On the three-dimensional vessels, I airbrush diluted 1:4 Klyr-Fyre before sifting. The wetted particles must dry before I fire. This layer of sifted transparents is fired to orange peel at 1500°F. When they come out of the kiln, the switch covers are weighted with a press plate to minimize warping. The heat changes the vessel forms slightly during cooling, and I incorporate these changes while I create the piece.■

"Burnt Offering XII" vessel, 10"h x 7" diameter, Limoges sgraffito enamel on spun, cut, sewn, and riveted copper form, silver foil, satin finish. *Photo by Jonathan Wallen*

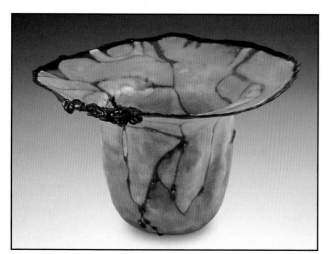

"Topography Cup 1" cup, 3"h x 4" x 3", Limoges sgraffito enamel on raised, electroformed copper, copper wire, silver foil, satin finish. *Photo by Barry Blau*

"Accrual Dish: Escaping Liquid" dish, 8.5" x 5.75" x 1" deep, Limoges sgraffito enamel on cut, riveted, and sewn copper, silver foil, satin finish. *Photo by Ralph Gabriner*

"Less Is More" vessel, 8"diameter x 3.25" deep, Limoges sgraffito enamel on spun, cut, and riveted copper form, silk-screened silver foil, satin finish. *Photo by Ralph Gabriner*

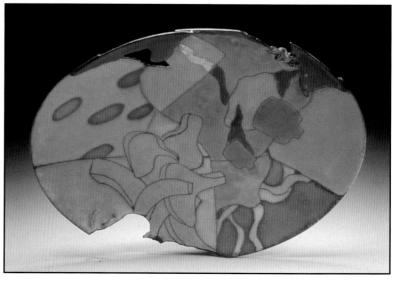

THE GALLERY
OF CONTEMPORARY ENAMELISTS

BARBARA ALTSTADT

Barbara Altstadt, a painter and a ceramicist, followed a natural path: the combination of kiln-fired enamels with her painting and ceramics. This combination makes her works very provocative and unusual. She never uses bright colors, but her pieces show a good balance of aesthetic values.

"Explosion" wall piece, 120 cm x 20 cm, sifted enamels on 23-ga copper pieces, assembled over iron. *Photo by Sergio Henrique V. Alberti*

"Pendulo I" and "Pendulo II" wall pieces, 64 cm x 12 cm and 80 cm x 12 cm, sifted enamels on 23 ga copper, assembled on cast iron. *Photo by Sergio Henrique V. Alberti*

JAN BAUM

Jan Baum earned her MFA, University of Massachusetts-Dartmouth and a BFA, Arcadia University. Her work is in the permanent collection of the Renwick Gallery, is included in The Penland Book of Jewelry and has appeared in Metalsmith. Recent exhibitions include: Museum of Decorative and Applied Arts, Moscow, Gallery Mukkumto, Seoul, and International Design Center, Nagoya, Japan.

Jan Baum

"Hot Style" locket 2.5625"x 1.188" Enamel on fabricated, die formed copper, setting: sterling silver, 14k yellow gold. *Photo by Courtney Frisse*

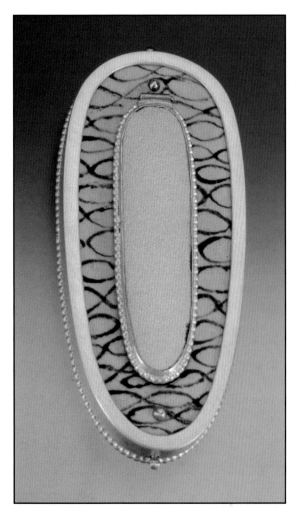

"Mark-ed" brooch, 3.0625" x 1.375", China paint on enameled copper, fabricated, setting: sterling silver, 14k yellow gold. *Photo by Bill Bachhuber*

MARLEEN B. BERG

Marleen B. Berg was born in 1957, Delft, Netherlands. Educated as a teacher, she trained youths to work with a variety of materials, including copper and enamel, which engaged her. She now makes her living with metalwork, enameling and graphic design. She sees her challenge to be to let metal texture and transparent enamels work together in one image. Marleen now lives in the inspiring historical town, Maastricht.

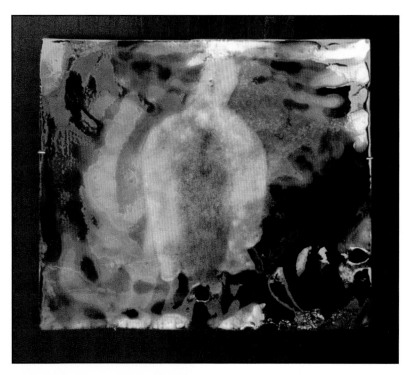

"And God Will Be With Them" panel, 6.3" x 6.3", final torch-fired enamel on written 28ga copper. *Photo by artist*

"Jeremiah" panel, 4" x 6.75", basse-taille enamel on 28 ga copper. *Photo by artist*

MIRTHES BERNARDES

Mirthes Bernardes was introduced to the art world through her ceramic sculptures. Her life turned around after attending enamel classes in 1993. Since then, she has dedicated her time to enameling and making the art of enamel known throughout Brazil through courses, workshops and the direction of the enamelists guild, NUBRAE. Her designs and enamels have won many prizes.

Mirthes Bernardes

"Flower Basket" wall piece, 100 cm x 80 cm, enamel sifted on 21 ga cut copper pieces and assembled. *Photo by Sergio Henrique V. Alberti*

"Pink Ipe" wall piece, 70 cm x 50 cm, enamel sifted on 21 ga, cut-copper pieces and assembled. *Photo by Sergio Henrique V. Alberti*

MARIA P. CARVALHO

Maria P. Carvalho was introduced to enameling at the Torpedo Factory Art League School in Alexandria, VA. Her background includes printmaking and many forms of jewelry that give her a wide-ranging aesthetic sensibility. She finds working with enamel is magical; the brilliance of color and the spontaneity of results are amazing. Her enamel work is a constant search. She combines motifs native to her homeland, Brazil.

Maria DaPenha R. Carvalho

"Patchwork" wall piece, 12.375" x 5.125", sifted, wet-packed enamel, leaf printing on 21 ga copper. *Photo by Sergio Henrique V. Alberti*

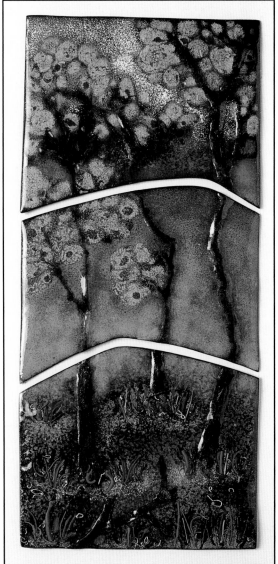

"Cerrrado Em Festa" wall piece, 5.125" x 10.75", sifted enamel, enamel threads and lumps. *Photo by Wagner Wada*

KATE CATHEY

Kate Cathey, a metal artist, has always been fascinated by color. She uses enamel to create rich and tactile surfaces that evoke our response to color. The metal object becomes the stage where layers of colored glass perform to dance in the presence of light.

Stickpins, 5"h x 2"w x 2"d, sifted and liquid enamel on copper. *Photo by Robert Diamante*

Rings, 1.25"h x 1"e x 1"d, sifted and liquid enamel on copper, setting: sterling, fine silver bezel, 14k gold granules. *Photo by Robert Diamante*

TZU-JU CHEN

Tzu-Ju Chen, BFA, Jewelry and Metalsmithing, Rhode Island School of Design, learned the plique-à-jour enameling technique at Penland School of Arts and Crafts. Enameling on three-dimensional forms, she means to ask questions on the perception of light in forms through colors. Her jewelry, one-of-a-kind pieces, is represented by the Nacy Yaw Gallery.

Tzu-ju Chen

"Scarlet Milonga" necklace, 16", plique-à-jour enamel on depletion gilded sterling silver, garnets, rubies, pink tourmalines. (detail) *Photo by Marty Doyle.*

"F.R. Venice" earrings, 3.5", plique-à-jour enamel, 18k yellow gold, moonstones, diamond roundels. *Photo by Marty Doyle.*

Karen L. Cohen

KAREN L. COHEN

Karen L. Cohen, enamelist and jeweler, started out as a hobbyist while working in development at AT&T Bell Laboratories. She studied at the School of Visual arts in New York City and then in private classes. Cohen specializes in cloisonné because of a love of the process. She mainly uses transparent leaded enamels for their luminosity as well as the depth and variety of their color.

"Paisley Pleasures" brooch, 2.25", cloisonné enamel on fine silver, 24k gold and fine silver foil, liquid gold overglaze, fine silver granulation balls, setting: sterling silver with fused silver design, 22k gold bezel, fresh water pearls. *Photo by Ralph Gabriner*

"Odotoglossum Rossi Orchid" wall piece, 8.75" square, cloisonné/basse-taille enamel on fine silver, 24k gold and fine silver wires, leaded enamels, fine silver balls. *Photo by Jack DeGeus*

NICOELLE DANIELLE COHEN

Nicoelle Danielle Cohen, raised in Coral Springs, Florida, took her first jewelry design course at Sante Fe Community College. There she studied enameling with Patricia Telesco. In May 2004, she received her BFA, Pratt Institute in Sculpture/Concentration in jewelry design; she studied with Patricia Madeja, Marybeth Rozkewicz and David Butler. She creates one-of-a-kind jewelry in her studio, Nicoelle Danielle Designs in Brooklyn, NY.

Nicoelle Danielle Cohen

"Kandinsky Muse Me" necklace, 1" x 3", wet-packed enamel on 18 ga copper, setting: sterling silver. *Photo by David Butler*

"Black Bird Fly" bracelet, champlevé enamel on depletion-gilded, etched sterling silver. *Photo by David Butler*

Cynthia Corio-Poli

CYNTHIA CORIO-POLI

Cynthia Corio-Poli finds enameling the perfect medium to pursue her love of color, glass and metal. By combining these elements with meaningful symbols, she produces unique and vibrant pieces. Her 20 years of many classes include those at the Gemological Institue of America and with Dorothea Stover at Montgomery College, Maryland. Enameling is her full time profession.

"Green Wave" ring, carved in wax, cast 18k yellow gold, enamel wet-packed, sapphire cabochon. *Photo by Eric Long*

Collection of Rings, carved in wax, cast 18k yellow gold and platinum, wet-packed enamel. *Photo by Ted Krohn.*

YVONNE CUPOLO

Yvonne Cupolo is a nationally known jewelry artist. She earned her MFA at State University of New York, Brockport. Her metalsmithing instructor was Arthur Palley. Self-taught in enameling, she has worked in enamels and precious metals for twenty-five years with inspiration from animals and organic forms. Her work evokes a personal response from those who share an affection for and fascination with the natural world.

Yvonne Cupolo

"Crow" pendant, 3" x 3.5", cloisonné enamel on fine silver, fine silver wires, setting: sterling silver, 14k gold, malachite, crystal.

"Coyote" brooch, 4" x 2.25", cloisonné enamel on copper, fine silver wires, setting: sterling silver, garnet.

MARILYN DRUIN

Internationally renowned enamelist Marilyn Druin was the recipient of numerous awards during her career. Her work is in the permanent collection of the Smithsonian's Renwick Gallery and the Newark Museum. Druin was a founding member and past president of the Enamel Guild NorthEast, a Trustee of The Enamel Society and played a key leadership role in developing enameling and jewelry workshops at the Newark Museum.

Marilyn Druin, 1996. *Photo by Mel Druin*

Necklace, 27", centerpiece: 1.5"w x 2.8"h, cloisonné/basse-taille enamel on fine silver, cloisons of 24k gold. Beads: cloisonné enamel on hand raised fine silver. *Collection of Carole Unger. Photo by Ralph Gabriner*

"Bird's Nest #3" sculpture (detail) in collaboration with Michael Good and Erica Druin, cloisonné/basse-taille eggs:1.5"h x 1"w, 24k gold cloisonné wires on fine silver, Nest: 3"h x 5"d, 22k gold and bronze patinated leaves on a stand of 22k gold, anticlastic formed by Michael Good. *Collection of Erica & Mel Druin. Photo by Bob Barrett*

DEE FONTANS

Dee Fontans is an internationally recognized enamelist, jeweler, performance artist, and fashion designer of Calgary, Alberta, Canada. She studied at New York's Parsons School of Design and SUNY, New Paltz. Exhibits and publication of her work have been in the United States, Germany, The Caribbean, and Canada. She teaches at Alberta College of Art and Design, Canada.

Dee Fontans

"Moon Drop" earrings, 1.95" x 1.17", enamel on copper, gold foil, bezels fused into the enamel, moonstones set in fine silver. *Photo by Charles Lewton–Brain*

"The Bath" placque, 5.85" x 8.19", enamel drawing on copper. *Photo by Charles Lewton-Brain*

JAIME FRECHETTE

Jaime Frechette graduated from Kent State University in 1993 with a BFA in Metalsmithing and Enameling. She studied with distinguished artists, Mel Someroski, Deanna Rob and Bruce Metcalf. Incorporating classical techniques with unconventional methods of her own has won her numerous awards.

Jaime W. Frechette

"Cavern Torch Flower" wall hanging, 14" x 7" x 5", hand-woven copper foil, formed and enameled. Insert: electroformed copper and nickel plated. The shades are of enameled copper mesh cloth, LED lights for the shades. *Photo by Andrea Millett*

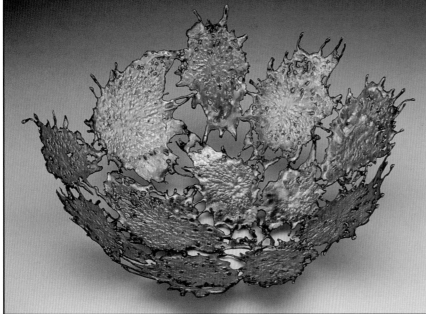

"Splatter" vessel, 13.5" x 12" x 6", sifted enamel on recycled, coiled, welded, hammered copper wire. *Photo by Mel Mittermiller*

DAVID FREDA

David Freda is an artist, jeweler, silversmith, educator, and avid naturalist. He is known for his experiments in hollow core casting and the development of a technique for firing high temperature glass enamels onto aluminum. He recently won the Grand Prize in the Saul Bell Design Awards Competition. His work is in many private and public collections including Victoria and Albert Museum, London; Mingei International Museum, San Diego; Museum of Art and Design, New York City.

David C. Freda with a gopher tortoise.

"The Mushroom Connoisseur" brooch, 4" x 1" x 1", wet-packed and sifted enamel on hollow core cast, hammer textured fine silver, 24k gold granulation, trim and findings 18k and 14k yellow gold, pearls. *Photo by Harold and Erica Van Pelt*

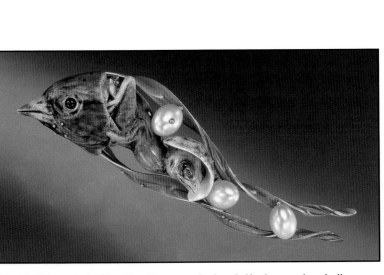

"Firebird" brooch, 3.25" x 1" x 1", wet-packed and sifted enamel on hollow core cast, hammer textured fine silver, 24k gold granulation, trim and finishing of 18k and 14k yellow gold, pearls. *Photo by Harold and Erica Van Pelt*

CID FREITAS

Cid Freitas constructs her enamels so that the theme and form are full of chromatic tones of luminosity. She approaches a piece with an awareness of the subject, the balance of volumes and the distribution of the elements like a puzzle. Synthesizing these points is essential for the innovative expression of this young artist.

Right:
"Balada II" wall piece, 14.25" x 8.125", sifted enamel on 18 ga copper assembled pieces. *Photo by Sergio Henrique V. Alberti*

Below:
"Polifonia" wall piece, 47.5" x 23.75", sifted enamel on 18-ga copper assembled pieces. *Photo by Sergio Henrique V. Alberti*

LENI FUHRMAN

Leni Fuhrman received a MFA from Pratt Institute. Trained as a painter and printmaker, she taught for many years at Kean University. When she branched out into metalsmithing, she became increasingly excited by the painterly possibilities inherent in enameling. She envisions enameling as "Nature in a Crucible," exploring in both her jewelry and sculptural objects the themes of the cataclysmic forces of Nature.

Leni Fuhrman

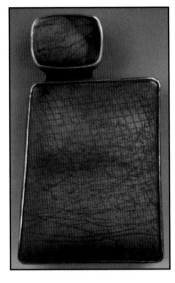

"Lava Series #2" brooch, 2.375" x 1.5" x .375" matte enamel on roller printed copper, spectralite, setting: fabricated fine and sterling silver, 22k gold. *Photo by Erik S. Lieber*.

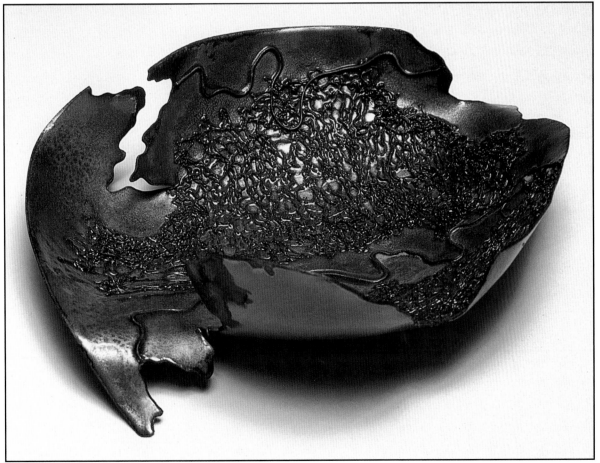

"Green Delta" sculpture, 7" x 6" x 2.25" raku enamel, sifted, wet-packed on copper, copper mesh and wire. *Photo by D. James Dee*.

MARIE ANGE HERCK GIAQUINTO

Maria Ange Herck Giaquinto, born in Brussels, Belgium in 1954, has had contact with the enamel world since her childhood. In Belgium, she studied with maitre-emailleur, Odette Gregoire. Marie now lives and works in Sao Paulo, Brazil. In 1977, she graduated Sao Paulo's School of Fine Arts. A large portion of her enamels is in very precise and unique cloisonné.

Marie Ange Herck Giaquinto. *Photo by Valerie Shaff*

"Pele" wall piece, 3.51" x 4.50" x 2.93", cloisonné spatulate, enamel on copper.

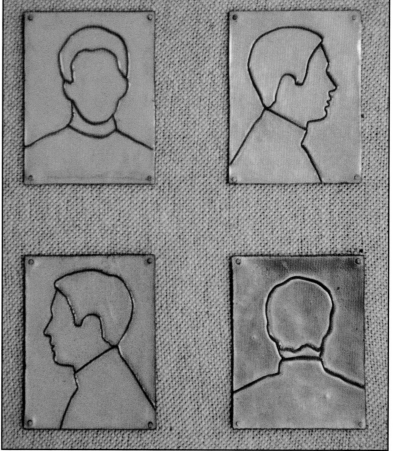

"Identity" wall piece, (4) 3.71" x 2.73", cloisonné, enamel on copper, wood and jute mounting.

NANCY GOODENOUGH

Nancy Goodenough, a renowned glass artist, was invited in 1997 to create an ornament for The White House. Her introduction to enameling was in a 1999 class with Sara Perkins. She approaches enameling from a glass perspective. She won the 2002 Niche Award for enameled metal and a finalist award in 2003. She was President of the Northern California Enamel Guild in 2003.

Nancy Goodenough torchfiring a bead.

"California Weather, or Not" vessel, 2.5" x 2" x 3", Limoges/cloisonné, wet-packed enamel on copper, fine silver wires, hand-raised metal. *Photo by Ralph Gabriner*

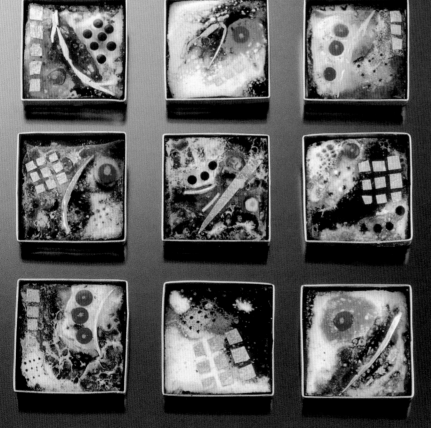

"Abstract Painting Series" brooch/pendants, 1.18" x 1.18 "x .25", Limoges/cloisonné enamel wet-packed on copper, 22k gold foil, fine silver wire. Frame: fabricated sterling silver. *Photo by Hap Sakwa*

Lisa Hawthorne

LISA HAWTHORNE

Lisa Hawthorne has been creating jewelry since the mid-seventies, working predominantly in sterling silver, gold and gemstones. She brought enamels into her work for more color and recently added diamonds into her enamels. She gathers ideas from diverse cultures as she travels. Her studio is in Coquille, Oregon, where she lives with her husband and daughter.

"Kool Kat" brooch, 2" x 1.75", cloisonné enamel on fine silver, roller printed hydraulic dye-pressed sterling, citrine. *Photo by George Post*

"Showtime" necklace, 18", enamel, 22k roller printed bimetal, diamonds, citrine, Australian opals, garnets, amethyst, peridot, black drusy, 22k gold bezels, 18k wires. *Photo by George Post*

LINDA HOFFMAN

Linda Hoffman is currently a Metals and Jewelry major at the University of Wisconsin-Milwaukee. She looks for used copper objects suitable for enameling at garage sales and re-sale shops. These discarded pieces are then transformed into new creations. She is influenced by artist, June Schwarcz, and by growing up in the 1970s.

Linda Hoffman

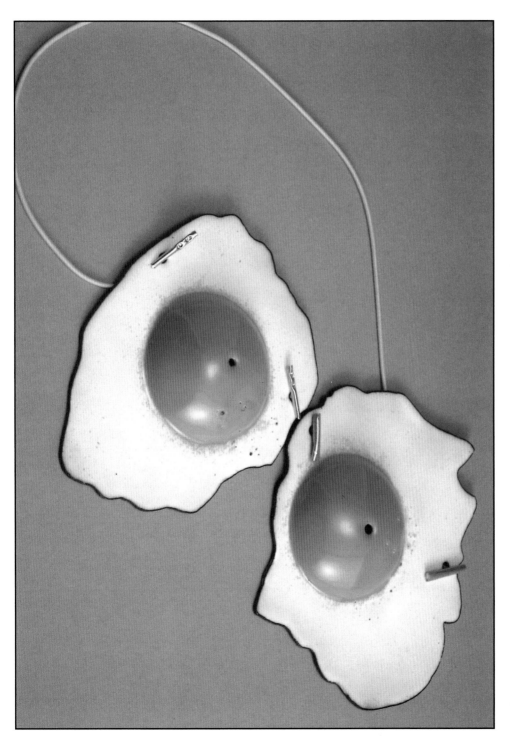

"Egg Mask" face ornament, 3.5" x 2.5", sifted, stenciled enamel on 18 ga copper.
Photo by Yevgenya Kaganovich

Sarah Hood

SARAH HOOD

Sarah Hood discovered a passion for jewelry making in California in the early 90s. She continued her study of metalsmithing at Parsons School of Design in New York City, where she began enameling. She simultaneously earned a BA in creative writing in 1997 from the New School for Social Research. In 1999, Sarah earned her BFA in Metal Design from the University of Washington.

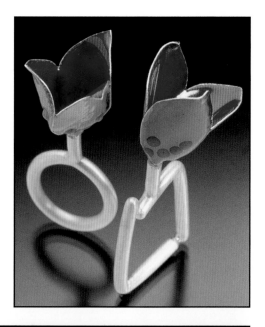

Two enamel rings, 1" x 2" x .75", sifted enamel on fine silver, glass beads. *Photo by Doug Yaple*

"Nilam" necklace (from Color Series #1: Dissemblance), 7" x 7"x .5", sifted enamel on fine silver, glass beads, sterling silver. *Photo by Doug Yaple*

ORIANA JARA

Oriana Jara is an artist who integrates enamel pieces with her ceramic sculptures. Born in Chile, she now resides in Sao Paulo, Brazil. In addition to her art work, Oriana develops social work projects to benefit Chilean immigrants. She has received many awards for her active voice against discrimination.

Oriana Jara

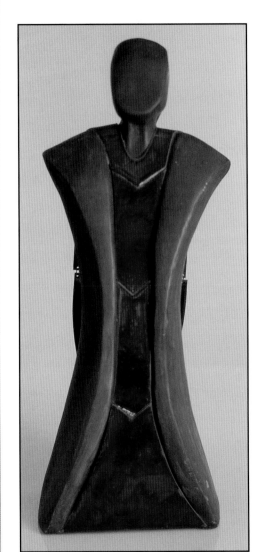

"Transparencia" wall piece with sculpture, 38 cm x 35 cm sifted enamel on 21 ga copper and acrylic sculpture. *Photo by Sergio Henrique V. Alberti*

"Vindima" ceramic sculpture, 6.375" x 5.125", 23 enamels on 21 ga copper inclusions. *Photo by Sergio Henrique V. Alberti*

VIVIAN KLINE

Vivian Kline has been an enamelist for 50 years, experimenting with all techniques and trying to push the enamel medium in new directions. She does not necessarily try to be different but often starts working with the question, "What if?" and so challenges herself to explore an idea in order to create something totally new.

Vivian Kline at work.

"Cityscape" necklace, 2" x 1.5" x 1", painted enamel and overglazes on copper. *Photo by Vivian Kline*

"Cityscape" five pins, 2" x 1.5" x 1", painted enamel and overglazes on copper. *Photo by Vivian Kline*

KATE KLINGENSMITH

Kate Klingensmith, growing up in a military family, lived and traveled throughout the United States. Her BFA from the Cleveland Institue of Art, was in enameling with a minor in silversmithing. She studied with John Paul Miller and Mary Ellen McDermott. Her work has been in many exhibitions. The landscapes of the West inspire her work.

Portrait of Kate Klingensmith.
Photo by Jeff Graef

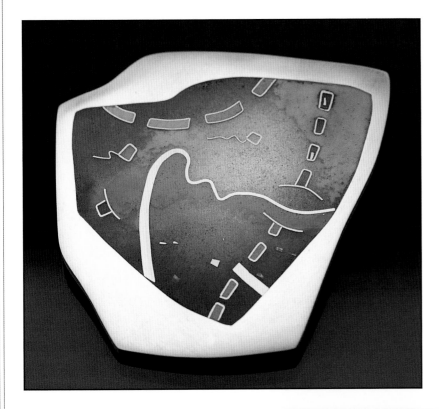

Brooch, 2"h x 2"w, champlevé/cloisonné enamel on fine silver, 24k gold wire, setting: fabricated sterling silver. *Photo by Steve Bigley*

Brooch, 1.5"h x 2"w, champlevé/cloisonné enamel on fine silver, 24k gold wire, setting: fabricated sterling silver. *Photo by Steve Bigley*

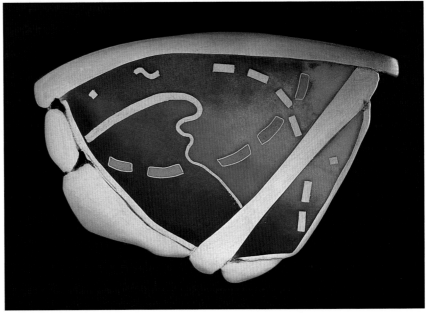

Susan Knopp in her studio.

SUSAN KNOPP

Susan Knopp, with a background in fine arts, approaches her cloisonné work as a painter. She also etches designs into her fine silver base, applies pigments and then wipes the base like an intaglio print. She feels that she is creating one-of-a-kind works of art rather than decorative jewelry.

"Fear Series #1" brooch, 2.25" x 2", cloisonné enamel on photo etched fine silver, foil, setting: sterling silver. *Photo by artist*

"Fear Series #2" brooch, 2" x 2", cloisonné enamel on photo-etched fine silver, foil, setting: sterling silver. *Photo by artist*

CHRISTINA LEMON

Christina Lemon is an artist, educator, and metalsmith who lives in Statesboro, Georgia. She received her BFA from Radford University and MFA from East Carolina University. Christina is an Associate Professor at Georgia Southern University where she teaches all levels of jewelry design, including enameling.

Christine Lemon

"Desert Series" pendant, 2" x 2.25" x .0625", cloisonné enamel on 22 ga fine silver, 24k gold wires, setting: sterling silver, 14k gold, citrine. Chain: sterling cable. *Photo by Seth Tice-Lewis*

"Ocean Series" earrings, 1" x .75" x .0625", cloisonné enamel on 22 ga fine silver, 24k gold wires, setting: sterling silver, 14k gold. *Photo by Seth Tice-Lewis*

Charles Lewton-Brain

CHARLES LEWTON-BRAIN

Charles Lewton-Brain studied and worked in Europe and North America. He lectures and publishes internationally on his research into rapid manipulation of metal and its surface for artistic and manufacturing purposes. He is Head of the Jewelry/Metals Program, Alberta College of Art and Design. He works in his studio, writes, and exhibits; he is President of the Canadian Crafts Federation

"Cage Pin B481" brooch, 5.5 cm x 7 cm, electroformed copper on stainless steel core, fusion gold inlay, enamel on copper, 24k gold heavy electrodeposit.

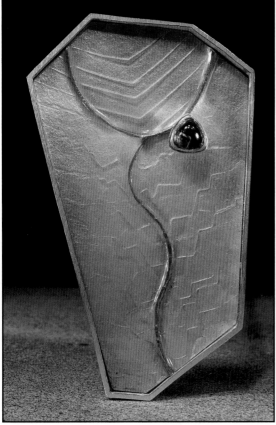

"Figure" brooch, 5.6 cm x 4 cm, fold forming, paper die printing, construction, enamel on copper, sterling silver, 24kgold, natural zircon.

KARLA MYA MAXWELL

Karla Mya Maxwell's life-long drive to create jewelry was directed toward cloisonné in 1976. This self-taught artist views enamel as structured spontaneity. The small format is challenging, but she is comfortable working in thousandths of an inch and views precision as a noble goal. The color, defining line work, permanence, and use of precious metals offer a possibility of expression that she finds compelling.

Karla Mya Maxwell

Pendant, 2.25" x 1.5", cloisonné enamel on fine silver, setting: sterling silver with matte onyx beads, 16.5". *Photo by artist.*

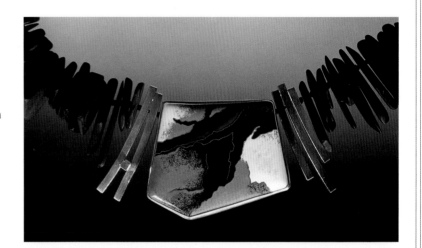

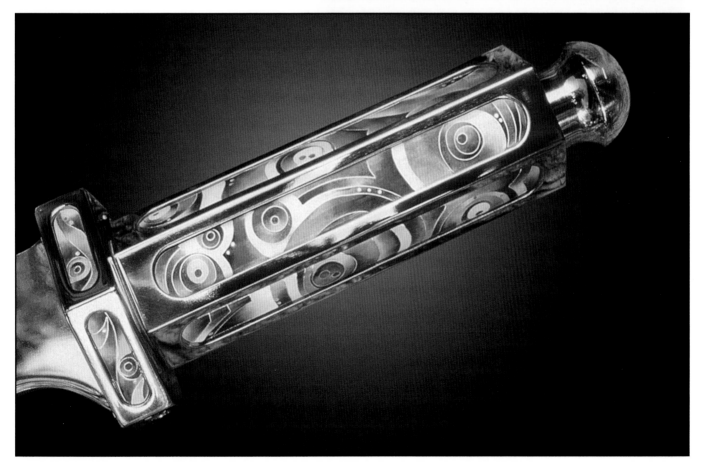

"Raindrops" cloisonné inserts (detail) of Leroy Remer's stainless steel knife, 12.5" x 2.25". *Photo by artist.*

MELS

Frequent travels have a deep influence on MELS' (Maria E. Lima) works. She uses different materials and vitreous enameled pieces to embellish and complete the ceramic masks in her new series. Enamel also has a special place on her iron sculptures.

"Ko Phi Phi" wall sculpture, 19 cm x 34 cm, sifted enamel on 23 ga copper inclusions on ceramics. *Photo by Sergio Henrique V. Alberti*

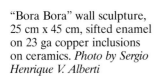

"Bora Bora" wall sculpture, 25 cm x 45 cm, sifted enamel on 23 ga copper inclusions on ceramics. *Photo by Sergio Henrique V. Alberti*

RENEE MENARD

Renee Menard finds the raw material and inspiration for her work from her love for travel, the natural world and photography. She uses selected aspects of the natural world, disassembles them and reconfigures them into images that are uniquely hers. Her enamels add punches of color to her pieces.

Vase, 4" x 12", sifted enamel on copper, kiln fired. *Photo by Helen Shirk*

"In Wendy's Garden" (detail) open bowl form, 8" x 14", sifted enamel on copper, kiln fired. *Photo by Helen Shirk*

Taweesak Molsawat

TAWEESAK MOLSAWAT

Taweesak Molsawat, Asst. Professor, Jewelry/Metals program, San Diego State University, holds an MFA with Honors, University of Kansas, Lawrence. His work, featured nationally and internationally, includes Victoria and Albert Museum, London and SOFA Chicago. Awards and honors are the SNAG Cultural Diversity Award and Best of Show, The Octagon Center for the Arts.

"Time Series: Under the Skin" brooch, 4.25" x 2.25" x 1.625", sgraffito enamel on copper, sterling silver, wooden ruler. *Photo by artist*

"Time Series: The Condition of Being a Human" brooch, 3" x 3"x .5", opaque enamel on copper, sterling silver, Lazertran, salt, red pepper, found objects. *Photo by artist*

KATHRYN OSGOOD

Kathryn Osgood is a jeweler, metalsmith and enamelist. Although the majority of her work is jewelry, she also creates hollowware. Her work explores the inherent malleability of metal to form organic shapes and interpret natural forms. Enameling adds an expressive element of color to her pieces. She maintains a studio in Greenville, North Carolina.

Kathryn Osgood

"Pod" necklace (open view), 4" x 1.5" x 1.5", sifted/wet-packed enamel on fabricated copper, setting: sterling silver, wood, brass, paint. *Photo by Robert Diamante*

"Flower Rings," 2" x 1" x 1" each, sifted/wet-packed enamel on fabricated copper, setting: sterling silver. *Photo by Robert Diamante*

DEBBIE PARENT

Debbie Parent has been working in enamels and metals for approximately seven years. She took her first enameling class in 1997 with Donna Buchwald and began exploring metal work in 1999 by taking a workshop with John Cogswell. Her other instructors include Ricky Frank, Merry-Lee Rae, Jean Stark, Mary Chuduk and Diane Almeyda. She is a member of The Enamel Guild South and the Enamelist Society.

Debbie Parent

"Night Falcon" pendant, 3.125" x 3",
fusing, cloisonné enamel on fine silver,
Ammonite. *Photo by Ralph Gabriner*

"Freedom" plaque,
16" x 8.25",
Limoges enamel on
copper

NATALYA PINCHUK

Natalya Pinchuk is originally from Russia. She now maintains a studio in Champaign, IL. Her work is exhibited nationally, including SOFA Chicago and New York with Charon Kransen Arts. The cover for the 2004 Exhibition in Print of Metalsmith *featured her work. She participated in the 2004 symposium on Enameling at Khors Gallery in Rostov, Russia.*

Brooch, 3.25" x 2.25" x 1.25", sifted and wet-packed enamel on electroformed copper; sterling silver wire holds plastic flowers in place.

Brooch, 2.5" x 2.5" x 2", sifted and wet-packed enamel on electroformed copper; sterling silver wire holds grass in place.

MARY RAYMOND

Mary Raymond, born in Sydney, Australia, studied at the School of Colour and Design. She learned a variety of enameling techniques at workshops in Australia, the United States and the United Kingdom. An enamelist for over 20 years, her work has been exhibited in many countries. An Associate of the Royal Photographic Society, Mary uses her travel photographs as inspiration for her designs.

Mary Raymond

Right:
"Sails" mural commissioned for Barrier Reef hotel, 10' x 13', stencils and sifted enamels on copper, separately enameled shim sails fused in final firing, transparent and opaque powder. *Photo by Arch Raymond, FRPS*

Below:
"The Escarpment" mural commissioned for hospital foyer, 15' x 60', stencils and sifted enamels on copper, transparent powder, liquid and lumps. *Photo by Arch Raymond, FRPS*

merry renk

Originally a painter, merry renk is a self-taught jeweler and enamelist. In 1947, she started making jewelry and became fascinated with enamels. An accidental happening while enameling led her to rediscover plique-à-jour. It fulfills an aspect of contemporary art because the enamel becomes structural and decorative. A grant from the National Endowment of the Arts aided her work in plique-à-jour.

merry renk

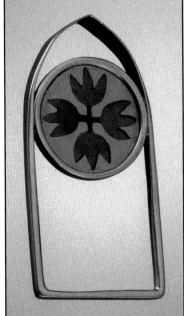

"Axis Mundi" ring, 2" x. 625" x. 125", plique-à-jour enamel, 14k white gold wires, 24k gold disc with plique-à-jour enamel. *Photo by Lee Fatheree*

"Once Upon a Time" hair comb, 6" x 1" x 8", plique-à-jour enamel, constructed of oxidized silver wire. *Photo by artist*

MARY S REYNOLDS

Mary Reynolds made the transition from painting to enameling upon seeing a glorious display of enamel test strips in the Metal Lab at Florida State University in the early 1980s. Her passion for rich color found a perfect outlet in the art of transparent enamels over precious metal foils. Recent work is jewelry with PMC and enamels.

Self-portrait of Mary S. Reynolds, 4" x 4", cloisonné enamel on etched copper, fine silver wires. *Photo by Capital Photo*

"Cat in Corn" plaque, 4.5" x 6.5", cloisonné enamel on etched copper, fine silver wires. *Collection of Pam & Vince Pahl. Photo by Capital Photo*

"Kaleidoscope Study" plaque, 4" x 4.5", cloisonné on etched copper, gold and silver foils, copper and fine silver wires. *Collection of Fae and Bud Mellichamp. Photo by Capital Photo*

D. X. ROSS

D. X. Ross has been drawing all her life. She learned grisaille enameling from Bill Helwig while pursuing her BFA in Printmaking. Self-taught, she set enamels into simple bezels until she enrolled for her MFA in metalsmithing at Tyler School of Art. Her work is in private collections, the National Gallery, Washington, DC and the Oakland Museum, CA. She is a popular teacher of enameling and metalsmithing.

"Surprised Bird" brooch, bird 1.5" x 1.5", china paints on base-coated enamel on 22 ga copper, setting: sterling silver, 22k gold bezel, sunstone. *Photo by D. X. Ross*

"Fish" unset enamel, 3.5" x 9", camieu and oxide enamel on 20 ga copper and gold foil. *Photo by D.X. Ross*

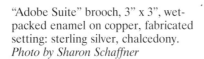

Sharon Schaffner

SHARON SCHAFFNER

Sharon Schaffner remembers enameling with her family around the kitchen table at age ten. "You've come a long way, Baby" is as true as ever. Professionally, she created enamel jewelry in the traditional cloisonné technique until she evolved to using a spontaneous painterly method. Now each layer builds until the feeling is as though looking deep into a clear lake.

"Adobe Suite" brooch, 3" x 3", wet-packed enamel on copper, fabricated setting: sterling silver, chalcedony. *Photo by Sharon Schaffner*

"Hanging Lake" brooch/pendant, 3.25" x 2", wet-packed enamel on copper in fabricated 12k gold-filled setting with lapis and garnets. *Photo by Gifford Ewing*

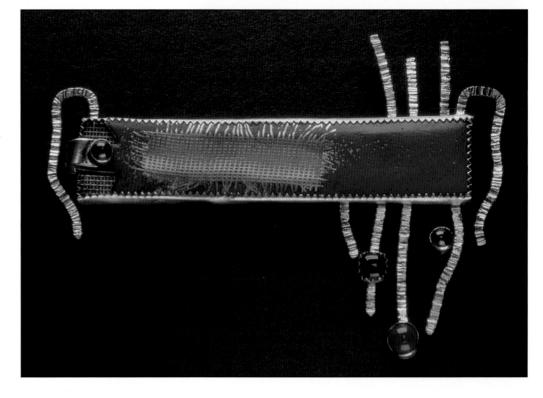

SYDNEY SCHERR

Sydney Scherr has exhibited her jewelry and sculptural work throughout the United States and internationally. Her formal education includes a BFA, Kent State University, Enameling with Mel Someroski; MFA, Southern Illinois University, with Brent Kington and Richard Mawdsley; Penland with Bill Harper. Working in series allows her to develop concepts. Her work includes narrative jewelry, sculpture and spiritual talisman rattles.

"Growth Rhythms Talisman" rattle (with artist Sydney Scherr), 86" x 20" x 4", 11 cloisonné enamels on copper, fine silver and copper wires, fine silver and fine gold foils, setting: fabricated 22 ga sterling silver and fine silver bezels, cold connections, aventurine, garnet, hematite, turquoise, rutilated quartz, tiger eye, ruby cabochons, purple heart wood.

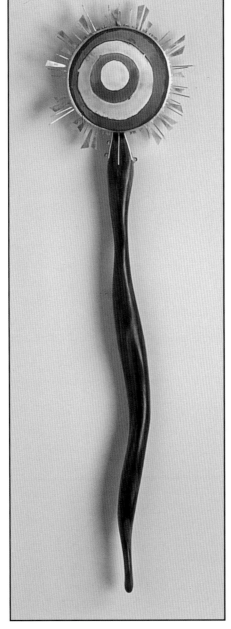

Far left:
"Lifelines" wall piece, 92" x 6" x 3", 14 cloisonné enamel panels on copper, fine and copper wires. Settings: fabricated 22 ga sterling silver and fine silver bezels, tiger eye cabochons, Bubinga hardwood.

Left:
"Sun Talisman" rattle, 32" x 8.5" x 3", cloisonné enamel on copper, fine silver and copper wires, setting: 22 ga sterling silver, fine silver bezel, carved ebony, amethyst, beach rocks rattlers.

Antonia H. Schwed

ANTONIA SCHWED

Antonia Schwed has been one of the leading jeweler/enamelists of New York City. She grew up in Bermuda and moved to New York at age 13. She was a well-known and beloved teacher at the Craft Student's League and the 92nd St Y. Her work is in many private collections around the world.

"Leopard" neckpiece, centerpiece 1.5" x 1.5", champlevé enamel on silver.

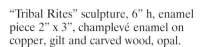

"Tribal Rites" sculpture, 6" h, enamel piece 2" x 3", champlevé enamel on copper, gilt and carved wood, opal.

DEBBIE SHEEZEL

Debbie Sheezel was first trained as an artist and then as a gold and silversmith. Her research of thirty years is evident in her versatility and virtuosity. Her work ranges from her mural, "Daintree," to unique one-of-a-kind pieces of jewelry and hollowware, which are collected and commissioned worldwide. She finds nature inspirational.

Debbie Sheezel

"Baubles" silver box, 3.5" x 2.5" x 3.25", cloisonné enamel on fine silver, silver and gold foils, 24k gold wire, box: sterling silver, bezel 22k gold, pearls.

"Daintree" mural, 60' x 24', enamel on copper, fine silver foils. Federa; Airports Commission for Brisbane International Airport.

KRISTIN MITSU SHIGA

Kristin Mitsu Shiga, metalsmith and enamelist, lives and works in Portland, Oregon. Kristin is a self-proclaimed tool junkie and curator of countless collections of inspirational artifacts. Her passion for making intricate objects is rivaled only by her passion for teaching others to do the same.

Kristin Mitsu Shiga

Detail of "Evolve" sculpture.
Photo by Courtney Frisse.

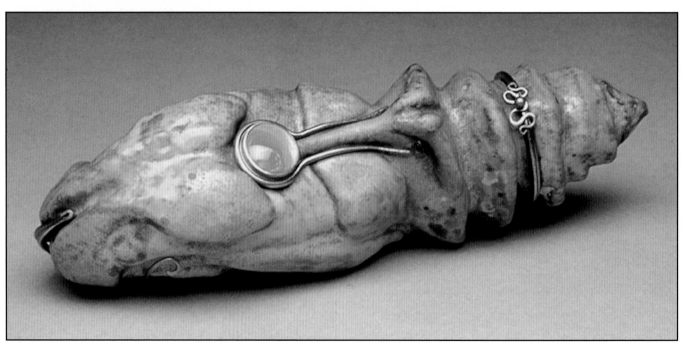

"Evolve" sculpture, 10"h x 2.75"w x 2.5"d, enamel on electroformed copper, sterling silver, pearls, lens, feathers, steel, worm.
Photo by Courtney Frisse.

HELEN SHIRK

Helen Shirk has been Professor of Art at San Diego State University since 1975. Her continuing interest in color has been reflected in her use of titanium, patina, colored pencils and, most recently, vitreous enamel. The vividness of opaque enamel enhances the sense of playfulness that can be achieved in small scale jewelry pieces.

Helen Shirk

Detail of "Whimsy" neckpiece.
Photo by Helen Shirk

"Whimsy" neckpiece, 8" x 8", sifted opaque enamel on 22 and 24 ga dark annealed steel wire that has been coiled into rosettes with stems. Rests on collarbone. *Photo by Helen Shirk*

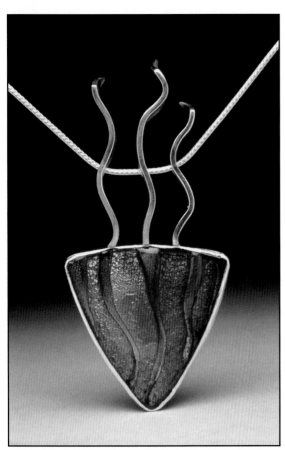

KORANNA SPURGEON

Koranna Spurgeon, although enameling since 1991, got a real start in 2001 when she was accepted into her first art fair, first gallery and began teaching at the arts center. In 2004, Koranna and Kristi Kloss conceptualized, curated and juried The Arsenal Metals and Enamels Exhibition.

"Angel Fire" pendant, 3.125" x 1.5", enamels on etched copper, setting: sterling silver. *Photo by H. Motto*

"Study In Negative Space" pierced vessel, 9" diameter x 6"h, sifted enamel on copper. *Photo by Koranna Spurgeon*

FELICIA SZORAD

Felicia Szorad, a practicing metalsmith with a 1998 MFA from East Carolina University, is head of the Jewelry and Metals program at Eastern Kentucky University. She has taught classes or workshops at Virginia Commonwealth University, Penland School of Crafts, Bowling Green State University and the Craft Students League. Felicia is a recent recipient of two Kentucky Arts Grants.

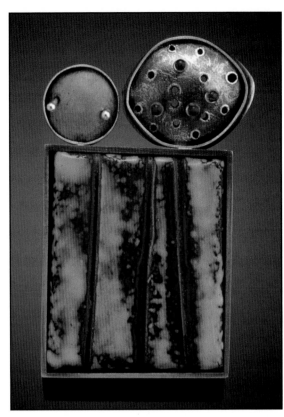

"Yellow Fold" brooch, 2.25" x 1.5" x .25", torch-fired sifted enamels on copper, setting: sterling silver. *Photo by Taylor Dabney*

Beaded bodice, 14" x 11" x 7.5", torch-fired enamel on copper, setting: sterling silver, gold-wrapped steel. *Photo by Taylor Dabney*

LEILA TAI

Leila Tai was born and raised in Beirut, Lebanon, during its "Golden Years." There she developed a passion for jewelry as an art form. She was educated in fine arts and metalwork at the American University of Beirut and the University of Wisconsin. Her most recent work is a collection of plique-à-jour butterflies and insects.

Leila Tai at her workbench.

"Blue Morpho" from Wings of Color collection, brooch/pendant, 1.75" x 1.5", pierced plique-à-jour enamel, 22k yellow gold, removable wings.

"Amberwing Dragonfly," from Wings of Color" collection, brooch, 3.25" x 2.5", pierced plique-à-jour enamel, body in fine silver and applied 22k gold, removable wings.

NOELLA TRIPPETTI

Noella Trippetti learned enameling by sharing her husband's enameling studio and his knowledge of cloisonné enameling. Her art training includes studies at University of New Hampshire, New Hampshire Institute of Art, The Sorbonne, and in Limoges, France. Noella is currently establishing an atelier in the Burgundy region of France.

"Morning Glories" wall piece, 8" x 10", cloisonné enamel on steel, fine silver wires. *Photo by Joseph Trippetti*

"Cheval" wall piece, 8" x 10", cloisonné enamel on steel, fine silver wires. *Photo by Joseph Trippetti*

179

Veleta Vancza

VELETA VANCZA

Veleta Vancza was born in Tucson, Arizona, 1973. She received her AAS, Fashion Institute of Technology; BFA, SUNY, New Paltz in 2000; MFA, Cranbrook Academy of Art in 2003. She is currently investigating the physical properties of vitreous enamel, specifically the ability for the material to significantly strengthen the metal substrate and its ability to be used as an adhesive.

"Homage To Form" wall piece (side view), 16" x 16" x 10", copper wire cloth, pieces hammered, thickly sifted enamel, fired, placed in frame and re-fired to unite the pieces. *Photo by artist*

White bowl, 11" x 11" x 5", sifted enamel on copper wire cloth. *Photo by artist*

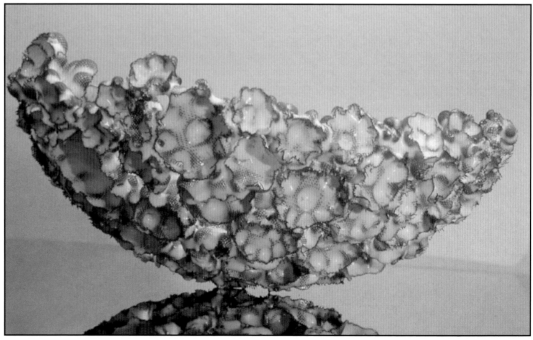

EMILY WATSON

Emily Watson, born in 1974, received her MFA from SUNY New Paltz in 2003. She uses her drawings to produce champlevé enamels that reference the body. The chains and strings of beads accentuate the drawn lines.

Emily Watson. *Photo by Dan Neuberger*

"Rendered Brooch #1," 3.5"l x 2.25"w x .125"h, champlevé enamel on 18 ga etched copper, constructed bezel with chain, sterling silver. *Photo by Dan Neuberger*

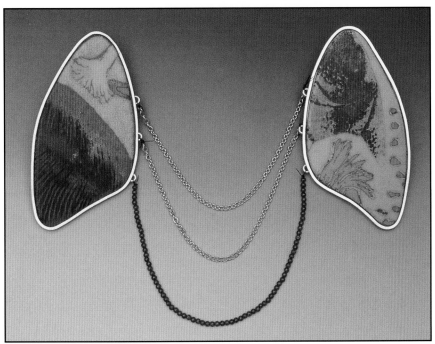

"Rendered Brooch #2," 3.6" x 6"w x .125"h champlevé enamel on 18 ga etched copper in constructed bezels with beads and chain. *Photo by Dan Neuberger*

DEBBIE WETMORE

Debbie Wetmore has been involved in the arts most of her life. With degrees in communication and commercial art, she has also studied enameling for the past fourteen years. Her primary interest has been in experimental techniques and in cloisonné. Her work is currently shown in Houston and Santa Fe and is included in the Smithsonian's collection at the Renwick Gallery.

Debbie Wetmore

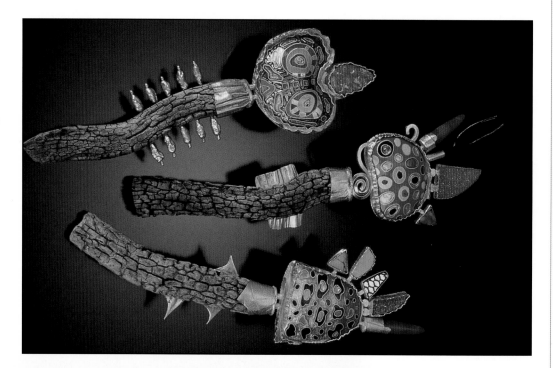

"Terra Incognita" brooches, 5" x 1.5" x .5", cloisonné, 24k gold wires on fine silver, 22k gold bezels, sterling silver, wood, coral, pearls, glass, assorted stones. *Photo by Robert Diamante*

"Crimson" necklace, 13" x 4" x 1", enamel on copper, sterling silver, street art caps, assorted stones. *Photo by Robert Diamante*

KAY WHITCOMB

Kay Whitcomb has sought both technical and aesthetic challenges, and her enamels since 1950 have won many international awards. She was graduated from the Rhode Island School of Design, jewelry major, in 1942 and apprenticed with Doris Hall for two years. She founded the Enamel Guild:West in 1976. She developed her architectual pyro enamel on steel technique in Belgium. A teacher, curator, juror, historian, innovator, Kay has been editor of "Cloisonné Collectors" since 1980.

Kay Whitcomb at her kiln.

"Portal of Sunset" pendant, gold plated, champlevé enamel on copper, cloisonné enamel beads, gold cloisons, sterling silver spacers.

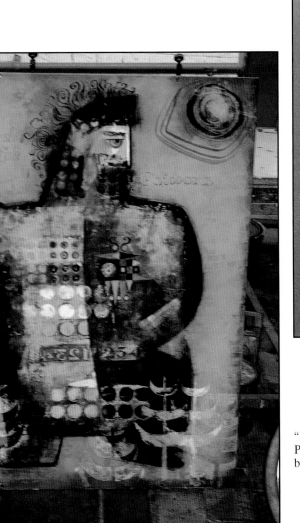

"Fisherman" wall piece, 42" x 33", Pyro enamel on steel, industrial oil-based screening pastes.

183

Teaching in the etching studio at Craft Students League, New York City.

KATHARINE S. WOOD

Katharine Wood, a second generation enamelist and daughter of Antonia Schwed of New York City, teaches enameling, which is her other passion, at the Newark Museum, 92nd St. Y, and the Craft Students League. She was co-creator with Paul Silverman of PNP transfer etching technique for enamelists and metalsmiths. Her work has focused on champlevé. She also loves combining techniques, mixing media, and working experimentally.

"Terra Incognita #2" wall piece, 14" x 14", champlevé enamel on copper. *Photo by Paul Silverman*

Paperweight, 2" x 6", multi-level champlevé/basse-taille enamel on copper.

SALLY WRIGHT

Sally Wright was led into enameling by her love of color and her training as a painter. Endless possibilities in the medium inspire her jewelry, sculpture and wall pieces. She is currently working on a series of intimate abstract landscapes. She exhibits internationally. Teaching venues include Newark Museum, Craft Students League and Museum of Art and Design, New York City.

Sally Wright

"Journey" pendant, 2.25" x 1.5", enamel on copper, wet inlay, stencils, firescale, palladium foil. Set in sterling silver, fine silver. *Photo by Ross Stout*

"Flatlands" wall piece, 12" square, champlevé enamel on copper, framed 24k gilt. *Photo by Ross Stout*

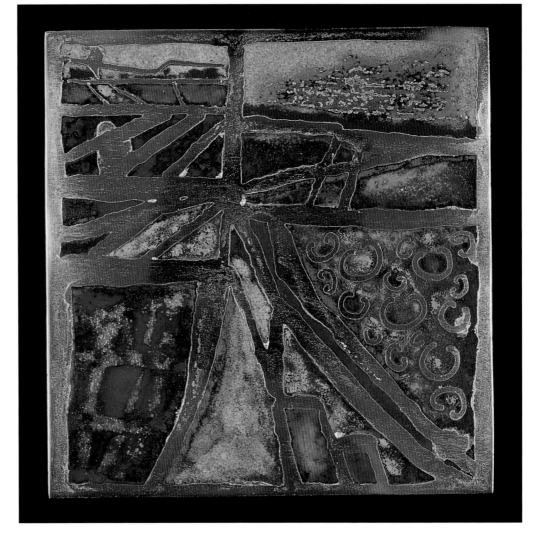

CHARTS

APPROXIMATE COLOR - TEMPERATURE INSIDE KILN

Dull red	650°C	1200°F
Warm red	750°C	1380°F
Cherry	800°C	1470°F
Bright red	900°C	1650°F
White	1280°C	2330°F

CERAMIC FIRING CONES

016	735°C	1357°F
015	770°C	1418°F
014	795°C	1463°F
013	825°C	1517°F

METAL MELTING POINT

Copper	1083°C	1980°F
Gold	1060°C	1945°F
Fine Silver	960°C	1760°F
Sterling	898°C	1640°F
Steel	350°C	2460°F

METAL GAUGES AND WEIGHTS

B & S	Inches	Millimeters	Drill Bit	Copper lb. Wt. sq. ft.	Sterling Silver sq. in. Troy oz.
29 ga	.012	.305	80		
26	.015	.381	79	.71	.087
24	.020	.508	76	.89	.110
22	.025	.635	72	1.13	.139
20	.032	.813	67	1.41	.175
18	.040	1.0126	59/60	1.79	.221
16	.050	1.270	55/56	2.25	.278
14	.064	1.625	52	3.84	.351

Avoirdupois Weight
28.35 gm = 1 ounce
16 drams = 1 ounce
16 ounces = 1 pound

Troy Weight
24 grains = 1 dwt.
20 pennyweights (dwt) = 1 ounce
12 ounces = 1 pound

Gram Weight
31.103 gm. = 1 troy oz.
  1.555 gm. = 1 troy dwt.

GUILDS & SOCIETIES

Argentina
Linea Libre
http://www.linealibre.com.ar/
Valeria Serrudo - Hugo Ostermann
Cuenca 3360 - 2 piso- departamento D
Capital Federal - Buenos Aires -
ARGENTINA - CP ABD1417C
Email: info@linealibre.com.ar

Australia
Australian Enamel Newsletter
c/o B. Ryman
71 George St.
Thirroul 2515
NSW Australia

The Enamellers Association
c/o Mrs. Ann Thomson.
10 Onthonna Terrace
UMINA NSW
Australia 2257

eNAMEL Online Newsletter
Allan Heywood, Editor
http://www.heywoodenamels.com/enews/

Belgium
Creativ-Kreis International - Belgium
Hedwig Luykx
Jan van Heelstraat 2B
3440 Zoutleeuw
Belgium
Email: hedwig@original-luykx.be

Brazil
Nubrae - Núcleo Brasileiro da Arte do
Esmalte
http://www.nubrae.com/
Mirthes Bernardes, President
Maria Carvalho, Cultural Director
Avenida Morumbi, 5594
São Paulo, SP
Brazil
Ph: (55) (11) 3063-9360
Email: presidencia@nubrae.com

Canada
The Canadian Enamellist Association
http://www3.sympatico.ca/c.blackburn/
enamel/
David Hustler, President
1 Ojibway Ave.
Toronto, Ont., M5J 2C9
Canada
Ph: 416-203-0962

France
GIRAEFE - Maison de l'Email
Michel Coignoux, President
171 Rue de la République B.P. 82
F-39403 Morez
France
Ph/Fax: 03 84 33 31 29
Email: jacques-clout@wanadoo.fr

Georgia, Republic of
Georgian Enamel Art
http://www.osgf.ge/chardingallery/
Ia Dvali
12 Sharden Str.
Tbilisi, 380005
Georgia
Email: ia_dvali@hotmail.com

Germany
Creativ-Kreis International
Gertrud Rittmann-Fischer
D-54534 Grosslittgen
Himmerod 4
Germany

Kunstverein Coburg E.V.
Kurt Neun
Hans-Holbein-Weg 10
D-96450 Coburg
Germany

Kunstlerwerkstaetten der Stadt Erfurt
Kulturdirektion
Lowetscher Strasse 42c
D-99089 Erfurt
Germany
Ph/Fax: 0361 735742

Great Britain
The Guild of Enamellers
http://www.guildofenamellers.org
Mr. Gerry Lomax, Secretary
Ullock Rise, Portinscale
N. Keswick, Cumbria CA12 5TX
Geoffrey Winter, Publicity Officer
Brighton Road, Lower Kingswood
Tadworth, Surrey KT20 6SX
England
Ph: 01737 830082

Society of British Enamellers
http://www.enamellers.org
Maureen Carswell, Secretary
Sainthill, Kingsland Road
Shrewsbury, Shropshire SY3 7AF
Ph:+44 (0) 1743-235-458
Email: maureencarswell@which.net
Ian Robertson - Chairman
Pat Johnson - Newsletter Editor
51 Webbs Road
London SW11 6RX
England
Email: pat.johnson@enamel.demon.co.uk

Hungary
Hungarian Enamel Center (est. 1975)
Nemzetközi Zománcmürészeti
Alkotómühely (Turi Endre, Director)
Bethlen krt. 16
H-6000 Kecskemét,
Hungary

Hungarian Society of Enamel Artists
(founded 1997)
Art Historian László Gyergyádesz, Jr.,
Chairman
H-6000 Kecskemét, Pázmány **P. u. 8.I/1**
Hungary
Ph: 76 417 378; 06-30 287 2549
Email: gyergyadeszzsu@axelero.hu

India
The India Enamel Society
c/o Veenu Shah
B-25, Chiragh Enclave

New Delhi - 110048 India
Email: VeenuS@nde.vsnl.net.in

Israel
Enamel Section, Israel Designer
Craftsmen's Association
Marga Michaeli, Coordinator
P.O. Box 17087
Tel-Aviv 61170
Israel

Italy
Creativ-Kreis International - Italy
Mrs. Miranda Rognoni, President
Via Kennedy, 4
20048 Carate Brianza
Milan, Italy
Ph/Fax: 0362-905972

Japan
Japan Enamelling Artist Association
6F-A Asakawa Bldg. 1-19-13
Hyakunin-cho, Shinjyuku-ku
Tokyo, Japan, 169
Ph: 03-367-3587

Japan Shippo Conference
Yohko Yoshimura
Kanda-Nishikicho Bldg. 403
3-20 Kanda-Nishikicho
Chiyoda-Ku Tokyo, Japan
Ph/Fax: 03-3219-7805

Mexico
Asociacion Mexicana de Esmaltistas
A.C. (est. 1991) (Association of Mexi-
can Enamelists)
Norma E. de Guevara, President
Bosque de Olivos #245
Bosque de las Lomas
México D.F. 11700
Fax: (52) 55-55-96-29-70
or (52) 55-52-51-75-32
Email: norel@att.net.mx

Netherlands
Society of Dutch Enamellers (est. 1983)
http://www.enamellers.nl
Gré Dubbeldam, President
De Galop 15, 8252 Dronten,
Netherlands
Ph/fax: +31-321-313.661,
Email voorzitter@enamellers.nl
Ruud Kaper, Secretary
J.G. Heuthorststraat 24, 7009 Doetinchem
Netherlands
Ph/fax: +31-314-394 804
Email: secretaris@enamellers.nl

Russia
Russian Center International
RCJ "Emalis"
Alexandr Karikh
Ulitsa Swobody, House 56, Whg. 38
Jaroslawl 150014
Russia

Stawropolskiy Krai
Nikolai Vdowkin
Ul. Juchnaya, N3
c. Pobegailowka, RUS-357222
Russia

Spain
Centre D' Informacio I
Difusio De L'Art
De L'Esmalt (CIDAE)
http://www.cidae.com
Citutat de Balaguer, 17 Llotja
Barcelona 08022
Spain

United States
Enamel Guild of Creative Arts Group
Suzanne Kustner
108 N. Baldwin Ave.
Sierra Madre, CA 91024
Ph: (626) 355-8350
Email: ZRXDOC@aol.com

Enamel Guild of New Jersey
Marian Slepian
5 Overlook Drive
Bridgewater, NJ 08807

Enamel Guild/North East
Sandra Kravitz, President
115 Willow St.
Roslyn, NY 11577
Ph. 516-621-5584
Email: alank111@juno.com
Lois Grebe, Secretary
9 Woodside Circle
Yarmouthport, MA 02675
Email: JL.grebe@comcast.net

Enamel Guild South, Inc. (est. 1975)
Marilyn Tendrich, President
16500 SW 74 Avenue
Miami, FL 33157
E-mail: msnabottle@boyssuits.com
Audrey B. Komrad, Newsletter Editor
5720 Maggiore Street
Coral Gables, FL 33146

Enamel Guild: West
Steve Artz, President

Jean Vormelker, Editor-*Vitreous Voice*
425 N. Shattuck Place
Orange, CA 92866-1232
E-mail: jean@jvormelker.com

The Enamelist Society, Inc.
http://www.enamelistsociety.org/
6105 Bay Hill Circle
Jamesville, NY 13078

Great Lakes Enamel Guild
8040 S. 66th St.
Franklin, WI 53132-9030
Ph: (414) 425-2465
Email: barb@barbpel.com

National Enamelist Guild (est. 1973)
Kathy Bransford, President
1802 Brookstone Ct.
Vienna, VA 22182

Northern California Enamel Guild
(est. 1975)
http://www.enamelguild.org/
Sandra E. Bradshaw, Newsletter Editor
Judy Stone, Corresponding Secretary
P.O. Box 254
El Cerrito, CA 94530
E-mail: jstone@cwnet.com

Northwest Enamelist Guild
c/o Ely E. Wilder, President
892 10th Court
La Fayette, OR 97127
Ph: 503-864-2403

Ohio Valley Enameling Guild
P.O. Box 310
Newport, KY 41072
Ph: 859-291-3800

San Diego Enamel Guild (est. 1981)
Karim Carlock, President
Studio 5 - Spanish Village Art Center
Balboa Park, CA 92101

SNAG (Society of North American
Goldsmiths) publication, *Metalsmith*
1300 Iroquois Avenue #160
Napierville, IL 60563
Ph: 630-778-6385
www.snagmetalsmith.org

Palm Beach Enamel Guild
Margery Prilik, President
89 Seville 'F'
Delray Beach, FL 33446
Ph: (561) 638-3477

A.R. Levine, Communications Contact
867 Lakeside Drive
North Palm Beach, FL 33408

Washington State Enamelists
c/o Coral Shaffer
1022 NE 68th St.
Seattle, WA 98115
Ph: (206) 525-9271

The Woodrow W. Carpenter Enamel
Foundation
P.O. Box 310
Newport, KY 41072
Ph: (859) 291-3800

SUPPLIERS

COPPER
Passaic Metal & Building Supplies, Inc.
5 Central Ave.
Clifton, NJ 07015-1849,
Fax: (973) 546-7179

Metalliferous
34 West 46th St.
New York, NY 10036
Ph: (212) 944-0909

American Metalcraft
2074 George St.
Melrose Park, IL 60160-1515
Ph: (800) 333-9133

American Copper & Brass
Oakland, CA

REFINERS
Hoover & Strong, Inc
Richmond, VA.
Ph: (800) 759-9997

Hauser & Miller Co.
10950 Lin-Valle Dr.
St. Louis, MO 63123
Ph: (800) 462-7447

Handy & Harman Co. (gold)
525 Nuber Ave.
Mt. Vernon, NY 10550

JEWELRY TOOLS, METALS AND FINDINGS
Myron Tobak
25 West 47th St.
New York, NY 10036
Ph: (800) 223-7550

W. R. Cobb Company
850 Wellington Ave.
Cranston, RI 02910
Ph: (800) 428-0040

Frei & Borel, P.O. Box 796, 126 Second
St., Oakland, CA 94604, (800) 772-3456

Rio Grande
6901 Washington NE
Albuquerque, NM 87109
Ph: (800) 545-6566
www.riogrande.com

Indian Jewelers Supply Co.
601 East Coal Ave.
Gallup, NM 87301
Ph: (505) 722-4451

Joseph P. Stachura Company, Inc.
435 Quaker Highway
Uxbridge, MA 01569
Ph: (508) 278-6525

T.B. Hagstog & Son
709 Sansom St.
Philadelphia, PA 19106, (800) 922-1006

Allcraft
135 West 29th St., Suite 402
New York, NY 10001
Ph: (800) 645-7124

Salvadore Tool & Findings, Inc.
24 Althea Street
Providence, RI 02907
Ph: (401) 272-4100

ENAMELS, ETC.
Thompson Enamel Inc.
publication *Glass on Metal*
PO Box 310
Newport, KY 41072
Ph: (800) 545-2776

Bovano of Chesire
(Soyer French Enamels)
800 South Main
Cheshire, CT
Ph: (800) 847-3192

JAPANESE NINOMAYA ENAMELS
Enamelwork Supply Co.
1022 N.E. 68th St.
Seattle, WA 98115
Ph: (800) 596-3247

The Enamel Emporium
1221 Campbell Road
Houston, TX 77055
Ph: (713) 984-0552

Leslie Ceramics
1212 San Pablo Ave.
Berkeley, CA 94706
Ph: (510) 524-7363

Schlaifer's Enameling Supplies
1441 Huntington Dr.
POB 1700
South Pasadena, CA 91030
Ph: (800) 525-5959
www.enameling.com

OVERGLAZES
Standard Ceramic Supply
PO Box 4435
Pittsburgh, PA 15205
Ph: (412) 923-1655

Thompson Enamel Inc.
PO Box 310
Newport, KY 41072

CHINA PAINTS
Maryland China Company Inc.
54 Main St.
Reisterstown, MD 21136-0307
Ph: (800) 638-3880

Rynne China Company
222 W. 8 Mile Rd.
Hazel Park, MI 48030
Ph: (800) 468-1987

TORCHES
Wale Apparatus Company Inc.
400 Front St.
PO Box D
Hellertown, PA 18055
Ph: (800) 334-9253

SAFETY GLASSES
Auralens
Ph: (800) 281-2872
www.auralens.com

PMC
PMC Guild
417 West Mountain Ave.
Fort Collins, CO 80521
Ph: (970) 419-5503

Rio Grande
7500 Bluewater Road NW
Albuquerque, NM 87121
Ph: (800) 545-6566

PMC Connection
3718 Cvalier
Garland, TX 75042
Ph: (866) 762-2529

ART CLAY
Art Clay USA, Inc.
2377 Cranshaw Blvd., Ste. 130
Torrance, CA 90501
Ph: (866) 381-0100

KILNS
Paragon Industries, Inc.
2011 South Town East Blvd.
Mesquite, TX 75149-1122
Ph: (800) 876-4328

KILN WIRES & REPAIRS
Steve Votta
Rhode Island
Ph: (401) 785-8334
Fax: (401) 785-8336

RISO SCREEN & SUPPLIES
Welsh Products, Inc.
932 Grant Ave.
Benicia, CA 94510
Ph: (800) 745-3255

Dick Blick, Art Materials
PO Box 1267
Galesburg, IL 61402
Ph: (800) 933-2542